KU-143-973

SAFETY AT WORK AND THE UNIONS

SAFETY AT WORK AND THE UNIONS

P.B. Beaumont

CROOM HELM
London & Canberra

Croom Helm Ltd, Provident House, Burrell Row,
Beckenham, Kent BR3 1AT

British Library Cataloguing in Publication Data

Beaumont, P.B.
Safety at work and the unions.
1. Industrial hygiene – Greag Britain
2. Industrial safety – Great Britain
I. Title
363.1'1'0941 HD7695

ISBN 0-7099-0097-X

Printed and bound in Great Britain by
Biddles Ltd, Guildford and King's Lynn

CONTENTS

TABLES

ACKNOWLEDGEMENTS

There are numerous people who provided information and advice which greatly assisted me in the preparation of this book. In this regard special mention must be made of George Bain and William Brown at the SSRC Industrial Relations Research Unit at the University of Warwick who allowed me access to their recent workplace industrial relations survey which made possible much of the analysis and results presented in chapters 3 and 6. This analysis of the workplace industrial relations survey was done in collaboration with David Deaton at Warwick to whom a special word of thanks is due for leading me through the murky waters of discriminant analysis. I am also extremely grateful to Jim Whyte and his colleagues at the Trade Union Studies Section of the Central College of Commerce in Glasgow, whose encouragement and assistance made possible the collection of the questionnaire data analysed in chapter 5. As always a special debt of gratitude is owed to Maureen Robb for her invaluable research assistance; scarcely a chapter in this book does not contain the results of her work. The many helpful discussions with safety representatives, safety officers, members of the Health and Safety Executive and my academic colleagues, in particular John Leopold and Rob Coyle, are also gratefully acknowledged. Finally, thanks goes to various secretaries at Glasgow University, especially Moira MacAskill, who have patiently grappled with different drafts of this book at various periods of time.

Chapter One

INTRODUCTION

The passage of the Health and Safety at Work Act 1974, specifically the regulations on safety representatives and safety committees which became law in October 1978, provides industrial relations scholars in Britain with an opportunity to make a major contribution to both the academic literature and public policy discussion on workplace health and safety. The fact that industrial relations scholars have not previously made such a contribution is, at least at first glance, rather surprising in view of the sort of figures set out below in Table 1.1.

Table 1.1: A Comparison of Days Lost Due to Industrial Disputes and Industrial Accidents, Great Britain, 1960-70

	Millions of Days Lost due to:-	
	(1)	(2)
Year	Industrial Disputes	Industrial Accidents
1960	3.0	21
1961	3.0	19
1962	5.8	20
1963	1.7	20
1964	2.0	22
1965	2.9	23
1966	2.4	24
1967	2.8	23
1968	4.7	23
1969	6.9	23
1970	11.0	23

Source: K G Lockyer, *Factory and Production Management*, Pitman, 3rd Edition, London, 1974, p.33.

The contents of Table 1.1 indicate that the number of days lost through industrial accidents was frequently 8 to 10 times the amount that was lost through strikes, while even in the higher strike years (1962, 1969, 1970) the figure for time lost through accidents was still more than double the amount lost through strikes. However, despite the fact that accidents are a far more important cause of time lost in industry, it has been the causes and consequences of strikes, rather than industrial accidents, which have overwhelmingly attracted the research attention of industrial relations scholars. Admittedly the study of industrial accidents has been a long-standing research interest of behavioural scientists, particularly of industrial psychologists, but many of the behavioural scientists who have studied industrial accidents would not consider themselves to be in the field of industrial relations. More importantly, the institutionalists who have for so long dominated the study of industrial relations in Britain have paid remarkably little attention to industrial accidents, much less to the more broadly defined subject area of workplace health and safety. The reason for the lack of attention on the part of the institutionalists would seem to derive in large measure from the relatively limited extent and nature of union involvement in the subject area of workplace health and safety in Britain prior to the passage of the 1974 Act. This point can be illustrated by considering the following schema of ways in which unions can, at least in principle, make a contribution to improving workplace health and safety:-

(i) through lobbying and supporting the passage of government legislation designed to regulate workplace health and safety conditions;

(ii) through aiding and representing workers in accident compensation claims;

(iii) through negotiating compensatory wage differentials for workers in high risk jobs and industries;

(iv) through pushing for the establishment of joint union-management health and safety committees;

(v) through negotiating safety provisions of a preventative and/or compensatory nature in collective bargaining agreements.

As we shall see, trade unions in Britain exerted some influence via all of the above means prior to the 1974 Act, but overwhelmingly their attention was concentrated on the first three means listed above. The priority attached by unions to essentially the compensatory aspects of workplace health and safety has been alleged to have deflected them from playing a more active and positive role in workplace health and safety preventative matters. The resulting imbalance of treatment was evidenced, according to Kinnersly, by contrasting the importance attached by unions to obtaining sizeable member compensation for accidents with the fact that only three out of 130 unions had a full-time safety officer, and only about 200 shop stewards a year attended the single one-week course on health and safety that was put on at the TUC training college in London in the early 1970s.(1)

The safety representative and safety committee provisions of the 1974 Act provide the opportunity for unions to redress the long-standing imbalance between the compensatory and preventative aspects of their involvement in workplace health and safety matters. These provisions of the 1974 Act have already been hailed as having important implications for extending industrial democracy. The essence of this line of argument is that the subject area of workplace health and safety, which has for so long been dominated by unilateral management decision-making at the individual workplace, and a framework of law that has taken a highly 'paternalistic' attitude towards the issue of employee and union involvement, is to become at least an area of extensive joint discussion, and possibly one of joint decision-making. The essence of this line of argument is similar to that which has been put forward in relation to the grievance procedure in the United States, namely that "... whether or not organised workers are successful in pursuing specific actions through the grievance procedure, it is the availability of that procedure and the sharing of decision-making authority with management which it represents that reflects the impact of unionism on personnel management."(2) This argument that the safety representative/committee provisions are important for what they symbolise and thus can be seen as an end in themselves (i.e. the industrial democracy dimension) can be taken a stage further by considering their potential contribution to improved workplace health and safety. The importance of this consideration follows from the fact that union members are disproportionately employed in industries with above average accident rates. This proposition can be illustrated by considering two different measures of the organised workforce (collective agreement coverage and closed shop coverage) in the manufacturing sector in Britain. It is these measures which are available at the 2 digit industry level for the year 1978 and are, together with the relevant reported industrial accident figures, set out in Table 1.2.

We computed Spearman (rank) correlation coefficients between columns (1) and (3) and (2) and (3) respectively. The results were as follows:-

(1) and (3) $r_s = .708$ (significant at the .01 level)

(2) and (3) $r_s = .705$ (significant at the .01 level)

These positively signed, highly significant coefficients indicate that the high(low) organised industries were very much the high(low) accident rate industries. That is, trade union members were overwhelmingly employed in the high accident rate industries in the manufacturing sector in Britain.(3) These findings substantially increase the potential importance of the argument that safety representatives/committees should make a contribution to improved workplace health and safety. They also have, as we shall see in Chapter 4, a bearing on at least one major criticism that has been made of the nature of the safety representative provisions, namely that such representatives can only be appointed by recognised trade unions.

If the unions take the opportunity provided by these regulations to substantially increase their involvement in the preventative aspects of workplace health and safety then one can expect to see a substantial number of studies of the workings and impact of safety representatives and safety committees during the decade of the eighties. In keeping with the strong institutionalist tradition of industrial relations research in Britain these are likely to be overwhelmingly case studies conducted in a relatively small number of plants; indeed the case study approach has reached the ultimate degree of disaggregation in Britain in recent years with a number of books being based on the results of a single plant investigation. These small scale, in-depth studies are typically argued

Table 1.2: Collective Agreement Coverage, Closed Shop Coverage and Industrial Accidents 1978.

Industry	Collective Agreement Coverage, (Male Manuals) 1978 (1)	Closed Shop Coverage, 1978 (2)	Incidence Rate per 100,000 at risk of total reported accidents (3)
Food, drink and tobacco	76.0	53.0	4770
Coal and petroleum products	87.4	67.0	6310
Chemicals and allied industries	78.4	42.0	4190
Metal Manufacture	88.0	62.0	6600
Mechanical engineering	75.9	52.0	3960
Instrument engineering	61.0	13.0	1560
Electrical engineering	76.3	35.0	2380
Shipbuilding and marine engineering	92.0	68.0	5620
Vehicles	85.6	57.0	3310
Metal goods not elsewhere specified	71.0	39.0	3700
Textiles	80.8	30.0	3010
Leather, leather goods and fur	65.3	15.0	2180
Clothing and footwear	63.0	28.0	950
Bricks, pottery, glass, cement, etc.	81.4	46.0	4800
Timber, furniture	71.5	39.0	3200
Paper, printing and publishing	83.6	79.0	2610
Other manufacturing industries	67.1	50.0	3230

Sources: New Earnings Survey, 1978, Table 203; Department of Employment Gazette, January 1980. The accident figures were provided by the Health and Safety Executive.

to yield valuable, qualitative insights into the nature of on-going industrial relations processes that simply cannot be generated by the quantitative analysis of large scale bodies of survey data; the latter being held to be capable of only identifying structural based relationships. In our view the relative advantages of case study and survey analysis have not been fully maximised in the existing body of industrial relations research in Britain because of the disproportionate concentration on the case study method. The result of this imbalance between the two methods of study is a body of industrial relations research that may be characterised in the following terms,(4)

> ... there has been no evolution from the rich description ... to the development of a somewhat more analytical approach, i.e. one that places the problems or issues of interest in a theoretical framework and employs research strategies to test and ultimately refine the theory. Thus, the cataloging and refinement of the concepts introduced by the institutionalists have not yet taken place.

The above quote was in fact taken from a review of industrial relations research in the United States, but as a criticism it applies even more forcefully in Britain where there has been much greater, indeed almost exclusive, concentration on the case study approach. This fact was recognised in a review paper by Bain and Clegg who argued that case studies should only be carried out when larger, more representative bodies of data have already been analysed with a view to identifying the broad, structural parameters of the subject area under investigation.(5) In other words, case studies are most valuable in the latter stages of a large-scale research project where they can provide the detailed type of information and knowledge that simply cannot be obtained through the quantitative analysis of large-scale bodies of survey data.

Accordingly, the basic purpose of this book is, as its title suggests, to provide an empirical overview of the subject area of union involvement in workplace health and safety in Britain. This is done by undertaking various pieces of analysis on a number of more aggregative sets of data than are typically utilised in industrial relations research in Britain. The value of this approach should be in providing (i)an overall, analytical framework that later, more detailed studies can potentially build on, and (ii) a set of results that will constitute a broader context into which the findings of these later more disaggregated studies can usefully be placed. This book should therefore be seen as a piece of work intermediate between the very general, 'guide line' type of books produced largely by lawyers on the details of the 1974 Act, a number of which have already been published, (6) and the small scale case studies of the workings of safety representatives and safety committees that one can expect to see over the course of the next few years. It is our general contention that this 'intermediate level' of analysis has all too frequently been neglected in industrial relations research in Britain; the obvious exception to this statement being George Bain's major study of white collar trade unionism. (7)

Outline of the Book

Turning to the individual chapters of the book, we present in Chapter 2 a review and analysis of the problem of industrial accidents that confronts the newly established safety representatives and safety committees in Britain. Some of the size and cost dimensions of the industrial accident problem in Britain will be highlighted, and then an attempt will be made to identify some of the major factors that appear to account for the variation in accident rates between industries. This examination of one of the leading 'empirical regularities' in the industrial accident statistics should enable us to identify some of the specific factors in the industrial environment that should be of concern to safety representatives and safety committees in seeking to carry out their preventative functions. Chapter 3 then considers the extent and nature of union involvement in workplace health and safety matters prior to the passage of the 1974 Act. In this task we utilise the schema of union involvement that was outlined earlier in this chapter. A major part of this chapter will be devoted to developing and testing a model designed to identify the characteristics of plants that had voluntarily established joint health and safety committees prior to the passage of the 1974 Act.

In Chapter 4 we briefly outline the case for government intervention in the area of workplace health and safety and then consider the nature of health and safety legislation prior to the 1974 Act. The particular focus of attention here will be on the constraints that this legislation placed in the way of any extensive union involvement in the preventative area of workplace health and safety. The relevant provisions of the Health and Safety at Work Act 1974 are then considered as providing the opportunity for unions to reorientate their attention from the compensatory to the preventative aspects of workplace health and safety. Some early evidence as to the extent of establishment of safety representatives will be presented in the concluding section of this chapter. Chapter 5 then utilises the returns from a questionnaire administered to a sample of safety representatives to consider a number of questions and issues that have been raised about the activities and likely impact of these newly appointed union representatives at the workplace. The central unifying theme considered in this chapter is the question of whether these representatives see themselves as primarily negotiating with or consulting with management.

Chapter 6 moves from considering the safety representative function to an examination of joint health and safety committees. A variety of data sources are used in this chapter to examine the influence of the 1974 Act in establishing new committees and reforming already existing ones in the sense of producing changes in their composition and functions. A framework for identifying various measures, and determinants, of the 'effectiveness' of these joint bodies is developed in the remainder of the chapter. The experience of union involvement in workplace health and safety in the United States is then examined in Chapter 7. This review emphasises developments and experiences since the passage of the Occupational Safety and Health Act of 1970 which, as with the 1974 Act in Britain, opened up the possibility of substantial union involvement in the subject area. The review of this experience should provide a number of useful research and public policy lessons and insights for Britain. Finally, in Chapter 8 we seek to redress some of the emphasis of earlier chapters by highlighting a number of neglected issues for further research.

NOTES

1. Patrick Kinnersly, _The Hazards of Work_, Pluto Press, London, 1973, p. 318-9.

2. David Lewin, "The Impact of Unionism on American Business: Evidence for an Assessment", _Columbia Journal of World Business_, Winter 1978, p.95.

3. For a fuller discussion of the nature of the data and results presented here see P B Beaumont, Robert Coyle and John Leopold, "Health, Safety and Industrial Democracy: Some Further Considerations", *Employee Relations*, Vol. 2, No. 3, 1980.

4. Thomas A Kochan, "Theory, Policy Evaluation and Methodology in Collective Bargaining Research", *Proceedings of the Industrial Relations Research Association*, Winter 1976, p. 243.

5. George Sayers Bain and H A Clegg, "A Strategy for Industrial Relations Research in Great Britain", *British Journal of Industrial Relations*, Vol. XII, No. 1, March 1974, p. 102.

6. See, for example, Richard Howells and Brenda Barrett, *The Health and Safety at Work Act: A Guide for Managers*, IPM, London, 1975.

7. George Sayers Bain, *The Growth of White Collar Unionism*, Oxford University Press, Oxford, 1970.

Chapter Two

THE PROBLEM OF INDUSTRIAL ACCIDENTS IN BRITAIN

In this chapter we examine various dimensions of the industrial accident problem that faces safety representatives and safety committees in their task of trying to improve the quality of health and safety at the workplace. The importance of this examination follows from the fact that variation in the accident potential of the working environment is likely to be one of the key factors accounting for differences in the attitudes and activities of safety representatives. Moreover, the ultimate test of the value of these new institutional arrangements at the workplace is likely to be whether they, *ceteris paribus*, make a significant contribution to reducing the level of industrial accidents, as well as other indicators or measures of the state of workplace health and safety.

The basic emphasis of the chapter is on reviewing certain size and cost dimensions of the industrial accident problem in Britain. Specifically, we consider the observable costs of accidents, the British industrial accident rate(s) in relation to those of other countries, and then examine any changes in the accident rate in Britain through time. The final section of the chapter reviews the existing body of literature on the causes of accidents in order to develop and test a model designed to explain inter-industry variation in accident rates. The implications of this exercise for studying the 'effectiveness' of safety representatives and safety committees will then be explored.

The obvious point to note about the contents of this chapter is that the problem of workplace health and safety is defined solely in terms of safety; the problem of occupational diseases and health is ignored. This approach may be justified, at least in the short term, by the relative immediacy of the industrial accident problem; in the sense that accidents cause far more days lost than occupational diseases. However, the real, long term challenge to these new institutional arrangements is arguably that of workplace health problems as an attempt is made to come to terms with the effects of years of heretofore undetected or unknown exposure to toxic substances. This point must be borne in mind when considering the contents of this chapter, and indeed will be returned to in our final chapter where we consider the future of these institutional arrangements in relation to workplace health and safety problems in Britain.

How Serious is the Industrial Accident Problem in Britain?

In attempting to answer this question one needs to put the industrial accident problem in some sort of context by, firstly, noting that ten times more deaths occur on both the roads and in the home than at the workplace. However, fatalities constitute only a relatively small proportion of all industrial accidents, and there are numerous individual statistics that can be cited in support of the contention that industrial accidents constitute a social and economic problem of quite substantial proportions in Britain. For example, Aikin and Reid noted in the early 1970s that, (1)

> Approximately twenty-one million working days a year are lost because of accidents at work, about two thousand people a year are killed in such accidents. The cost to the country has been estimated at between £220 and £550 million a year, and the number of days lost at about six times as great as those lost through strikes.

Moreover, the compensation (in the form of common law damages) for such accidents has been estimated to constitute a tax on British industry of some £60 million per annum.(2) These sorts of statistics should, however, be seen in the light of the Robens Committee's comment that the relative lack of systematic and comprehensive studies of the private and social costs of industrial accidents constitutes a serious constraint on informed public policy discussion of the subject.(3) The Robens Committee itself distinguished between the costs falling on the employer, the Exchequer and the total economic and social costs to the nation. A number of individual cost estimates were cited under each of these headings, but in relation to the estimates of the all important social costs of accidents the Robens Report stressed that "the validity of the bases of calculation were usually very much open to question, a fact which reflects the limitations of present knowledge and technique on this subject."(4)

The above comment should be seen in the light of both the conceptual and measurement difficulties facing any attempt to answer the question of whether Britain has a 'serious' industrial accident problem. Conceptually one can talk of a socially optimal level of accidents, which is the level achieved when any further reduction in accident rates can only be achieved at a cost greater than the estimated benefits.(5) The value of this concept is largely as an academic counterweight to public policy discussion which often tends to implicitly assume that whatever the prevention costs of accidents they must be less than the benefits of lower accident costs. However, it is a non-operational concept that cannot be quantified and thus cannot provide the ideal reference point for judging the seriousness of any country's industrial accident problem.

In addition to the absence of this ideal yardstick or reference point for comparison purposes, there are various measurement problems associated with the available accident statistics themselves in most countries. The major source of industrial accident statistics in Britain are those reported to the various inspectorates concerned with ensuring observance of the relevant health and safety regulations. The accidents that are legally required to be reported to the Factory inspectorate, for example, have long been only those that involve a fatality or absence from work for more than three days. These legally reportable accidents do not constitute anything like a full count of all the industrial accidents that occur. In their evidence to the Robens Committee the Department of Employment estimated that the ratio of non-reportable to legally notifiable accidents was of the order of 30 : 1.(6) Furthermore, not all legally notifiable accidents are in fact reported to the inspectorate; the Department of Employment estimated the shortfall in this regard to be around 15-25 per cent in the manufacturing sector, and around 25-40 per cent in the construction industry.(7) These figures suggest that changes in the degree of employer compliance with the requirement to notify accidents could 'distort' any 'real' trend in the reportable accident rate through time, while such distortion could be compounded by exogenous changes in the propensity of workers to take time off for an accident of given severity. The trend in such accidents through time has nevertheless been used as some sort of measure of the seriousness of a country's accident rate problem. However, before considering the experience in Britain in this regard we briefly examine a second approach that has sometimes been adopted in trying to answer this question of how serious is a country's accident problem, namely a comparison of one country's accident rate with that of other countries.

A Cross-Country Comparison of Accident Rates

There have been attempts by bodies like the International Labour Office to bring about a more standardised collection and presentation of industrial accident statistics between countries, but progress to date in this regard has been relatively limited. The International Labour Office views accident *frequency* rates, which are computed by dividing the number of accidents (x 1,000,000) which occur during any given period of time by the number of hours worked by all persons exposed to risk during that period of time, as a better measure of accident risk than accident *incidence* rates, on the grounds that they are not affected by differences in working hours between industries and/or countries. In practice, however, it is very largely the latter measure, which is computed by dividing the number of accidents (x 10,000) which occur during any given period of time by the average number of workers exposed to risk during that period of time, that is available for most countries. Furthermore, international comparisons of accident incidence rates are usually only for fatal accidents, so as to avoid the 'under-counting' problems and distortions which were mentioned above. It is a set of fatal accident incidence rates that are presented in Table 2.1.

The first, and most obvious, point to make about these figures is that the fatal accident rate in construction is very much higher than in the manufacturing industry sector in all of the countries; in Britain, for example, it is typically four to five times that in manufacturing, a fact that has come in for considerable comment in various annual reports of the Factory Inspectorate. In seeking to compare the figures for the different countries there are inevitably difficulties due to

Table 2.1: Incidence Rates of Fatal Accidents in Manufacturing and the Construction Industry for Selected Countries, 1972-76

Country	Code	1972	1973	1974	1975	1976
Great Britain	- M I/C	4	4	5	4	3
	- C I/C	19	22	16	18	15
France	- M II/C	12	10	10	-	-
	- C II/C	47	45	46	-	-
Federal Republic	- M II/A	18	17	16	16	-
of Germany	- C II/A	39	37	33	35	-
Irish	- M I/B	7	10	8	9	5
Republic	- C I/B	14	15	15	8	9
Italy	- M II/A	8	8	8	-	-
	- C II/A	55	51	62	-	-
Netherlands	- M I/A	4	4	4	4	-
	- C I/A	13	12	8	10	-
Sweden	- M II/D	4	3	3	-	-
	- C II/D	8	6	8	-	-
Canada	- M I/C	14	15	21	15	10
	- C I/C	90	96	121	96	75
USA	- M I/D	4	3	3	3	-
	- C I/D	23	13	16	16	-

Japan	- M I/D	3	3	2	2	1
	- C I/D	19	21	16	13	6

Source: Health and Safety Executive, Health and Safety Statistics 1976, HMSO, London, 1979, p. 42.

I Reported Accidents A = rates per 100,000 man-years of 300 days each
B = rates per 100,000 wage earners (average)
C = rates per 100,000 persons employed (average)
D = rates per 100 million man-hours worked

II Compensated Accidents with A, B, C, D as above.

differences in collection and presentation methods, but the Health and Safety Executive has argued that most of the countries listed (except for the D coded countries) are on an approximately comparable basis so far as the population at risk is concerned. On this basis, therefore, the position in Britain does not appear to compare unfavourably with that in other advanced, industrialised countries. In making this sort of judgment one must recognise that there will be differences in the industry mix of the countries' manufacturing sectors that will obviously be an important influence on the overall manufacturing accident rate because of substantial inter-industry variations in accident rates. The potential strength of this influence becomes most apparent when we find that the high(low) accident rate industries in one country tend by and large to be the high(low) accident rate industries in other countries. Some evidence in support of this proposition is presented in Table 2.2.

In spite of certain industry aggregation difficulties (which anyone with a working knowledge of the main order industry headings in Britain will readily appreciate from the list in Table 2.2) we find that fully five of our six coefficients attain varying degrees of statistical significance. Accordingly it would seem that there is a not inconsiderable amount of support for the belief that the high(low) accident rate industries in one country are very much the high(low) accident rate industries in other countries. The most obvious implication of these findings is, as suggested earlier, that a country's <u>overall</u> manufacturing industry accident rate will be highly sensitive to the distribution of the workforce between the various industries in that sector. That is, if the industrial structure of one country is such that a relatively high proportion of its total manufacturing workforce is employed in the iron and steel and woodworking industries, for example, then it is bound to have a much higher, overall accident rate than a country that has a relatively high proportion of its workforce in printing or textiles. Moreover, given the potential strength of this structural influence, together with the fact that accident risks in individual countries may not vary a great deal across countries, it would seem somewhat harsh to criticise the former country for its relatively poor overall accident rate performance. The other implications of this analysis will be discussed in later sections of this chapter.

Table 2.2: Individual Industry Accident Rates, Selected Countries, 1970

Industry Order	USA	New Zealand	Japan	Britain
Food and tobacco	20.35	61.65	8.59	38.2
Textile mill products	10.4	35.7	4.26	26.8
Apparel and other textiles	7.7	12.1	1.84	7.9
Timber, furniture, etc.	28.05	56.95	15.2	31.3
Paper, printing and publishing	12.8	26.15	7.0	24.8
Chemicals and allied	8.5	41.0	4.77	35.6
Petroleum and coal	11.3	32.1	3.99	54.8
Leather and leather products	15.2	29.0	7.85	18.4
Stone, clay and glass	23.8	63.8	9.93	54.2
Primary metal industry	11.9	59.3	6.50	78.9
Fabricated metal products	22.4	82.1	9.81	39.7
Machinery, except electrical	14.0	69.3	5.78	26.85
Electrical equipment and supplies	8.1	44.1	2.27	24.6
Transportation equipment	7.9	35.3	5.71	34.5
Miscellaneous manufacturing	17.2	36.4	5.93	34.6

The resulting Spearman (rank) correlation coefficients were as follows:-
USA + New Zealand = 0.625 (significant at the .05 level)
USA + Japan = 0.907 (significant at the .01 level)
USA + Britain = 0.475 (significant at the .10 level)
New Zealand + Japan = 0.507 (significant at the .10 level)
New Zealand + Britain = 0.539 (significant at the .05 level)
Japan + Britain = 0.332

The Accident Position Through Time in Britain

The other approach which has been used to consider the 'acceptability' of a country's accident rate(s) is to examine the direction and extent of its change through time. In this regard Phelps Brown, for example, noted that,(8)

> We have no trustworthy measure of the course of non-fatal accidents, because the standards of reporting will have changed, but that does not apply to the fatal ones, and these gave a death rate from accidents at work that in the early 1900s was double and more than double what it has become fifty years later. Out of each ten thousand at work in 1952-6, the factory operatives lost about one killed a year, the railwaymen less than four, the coalminers less than six, the seamen less than eleven. In 1902-6 the corresponding losses were rather more than two factory operatives, nearly eight railwaymen, more than twelve miners, and fifty seamen.

In Table 2.3 below we set out some more recent figures for the total number of fatal and non-fatal accidents reported under the provisions of the Factories Acts for the period 1950-75.

The obvious sub-period to stand out in this Table is 1963-69 where there was a year to year increase in the number of non-fatal accidents reported to the Factory Inspectorate. This adverse movement was predictably of considerable concern to the unions during these years, and indeed was an important influence on their demands for statutory based joint health and safety committees, which we shall discuss in Chapters 3 and 4. This increase in non-fatal reported accidents during the years 1963-9 was still present when we controlled for any changes in the base population at risk. The reported accident rates (per 1,000 employed) for these particular years were as follows:

1963	22
1964	28
1965	30
1966	30
1967	32
1968	35
1969	35

In attempting to account for this adverse movement the logical starting point is to consider a

Table 2.3: Fatal and Non-Fatal Accidents Notified in Britain under the Provisions of the Factories Act, 1950-75

Year	Total Number of Notified Accidents	
	(1)	(2)
	Fatal	Non-Fatal
1950	799	192,260
1951	828	182,616
1952	792	176,718
1953	744	180,893
1954	708	184,893
1955	703	187,700
1956	687	184,098
1957	651	174,062
1958	665	167,032
1959	598	173,473
1960	675	189,551
1961	669	191,848
1962	668	189,490
1963	610	203,659
1964	655	267,993
1965	627	293,090
1966	701	295,909
1967	564	303,452
1968	625	311,805
1969	649	321,741
1970	556	304,039
1971	525	268,275
1972	468	257,669
1973	549	271,969
1974	479	256,451
1975	427	242,713

Source: Compiled from various issues of the *Department of Employment Gazette*.

business cycle influence, as it has long been argued that industrial accidents are positively correlated with the level of economic activity, rising in the upswing and falling in the downswing of the cycle. The most detailed study along these lines is undoubtedly that by Robert Smith who sought to account for the year to year changes in the accident frequency rate in the United States manufacturing sector for the period 1948-69.(9) This attempt involved the construction of a model which incorporated a time trend, an earnings variable and three business cycle variables; overtime hours, the accession rate and the capacity utilisation rate. The results of this estimation exercise, which generally accorded well with his *a priori* hypotheses, led him to conclude that "... fatigue, inexperience, and use of surplus equipment all appear to be important determinants of the rise in injuries as an 'upswing' progresses and their subsequent fall in recessions."(10) A similar, if less comprehensive and detailed, study has been undertaken in Britain by Steele who sought to account for quarterly changes in reported accidents over the period 1965-71.(11) This was done by reference to an index of labour scarcity (the ratio of unfilled vacancies to the level of unemployment) and the extent of overtime hours worked. The overtime hours variable was the most powerful in his estimating equation, having a strong, positive relationship with the number of accidents that occurred. The strength of his finding for the overtime variable suggests the strong possibility that the 'high accident' years 1963-69 were years of relatively high overtime working. In order to provide some perspective on this matter we regressed the number of non-fatal accidents per 1,000 workers (IA) on average weekly overtime hours (OT) for the years 1950-68, with the unemployment rate (UN) and engagement rate (EN) included as other possible business cycle indicators. The following results (with standard errors in parentheses) were obtained:

$$\underset{}{IA} = \underset{(2.79)}{1.14C} + \underset{(0.01)}{0.60T} + \underset{(1.91)}{2.22UN} - \underset{(2.54)}{0.36EN} \quad (R^2 = 0.69)$$

This is essentially an illustrative exercise, rather than an attempt to build a comprehensive model designed to explain accident rate variation over this period of time, and hence too much should not be made of the results obtained. Nevertheless it is worth noting that the overtime variable does perform usefully as an indicator of a business cycle influence on accidents; the unemployment and engagement rate variables are nowhere near significance in the equation. Moreover, the suggestion of a high overtime concentration in the high accident sub-period 1963-68 was supported by an examination of the 'raw' overtime figures. These figures revealed that the mean (median) number of overtime hours per week for the sub-period 1963-68 was 315.7 (330.5) compared to 221.9 (228.2) hours for the full period 1950-68. Furthermore, a simple correlation of r = 0.78 was obtained between the extent of overtime working and a dummy variable for the years 1963-68. Finally, it is interesting to note that a similar rise occurred in the accident rate for the manufacturing sector in the United States over approximately the same period of time. This movement, which formed an important part of the background to the passage of the Occupational Safety and Health Act in 1970, has been investigated in a much more comprehensive examination (12) than that undertaken here, and the results revealed that no 'special' adverse forces had been at work during these years; the increase was simply the result of a strengthening of traditional influences on accident rates, most notably a continuing upswing in the business cycle.

An Analysis of Inter-Industry Variation in Accident Rates

In addition to industrial accidents being a cost of relative economic prosperity another well documented empirical regularity in the industrial accident statistics is that the observed inter-industry variation in accident rates is remarkably stable through time. That is, the high(low) accident rate industries in any one year are very much the high(low) accident rate industries in other years. In support of this statement we set out in Table 2.4 some accident rate statistics for the manufacturing industry orders, at the 2 digit industry level, in Britain for the years 1972-4 (average), 1975, 1976, 1977 and 1978.

The quite remarkable stability in the industry accident rate structure is revealed by the following Spearman (rank) correlation coefficients which we calculated:

1972-4 (average) and 1975 = 0.998
1972-4 (average) and 1976 = 0.990
1972-4 (average) and 1977 = 0.983
1972-4 (average) and 1978 = 0.983

This high degree of stability in industry rankings across time is even maintained when we disaggregated to the MLH or 3 digit industry level for the manufacturing sector; an rs = 0.964 being obtained between industrial accident rates in 1972-4 and 1976. This sort of stability is certainly not unique to Britain.(13) Moreover, as we saw earlier, there is evidence to suggest that the high(low) accident rate industries in one country are the high(low) accident rate industries in other countries. However, in spite of the former, well documented empirical regularity there appear to have been relatively few systematic attempts to account for inter-industry variation in accident rates;(14) the task that is

Table 2.4: Inter-Industry Variation in Accident Rates Reported under the Provisions of the Factories Act,* at the SIC Order Level, for Britain, 1972-74 (Average), 1975, 1976, 1977 and 1978

Industry Group	Incidence Rate per 100,000 at Risk of Total Reported Accidents				
	1972-74 (average)	1975	1976	1977	1978
Food, drink and tobacco	4520	4370	4480	4720	4770
Coal and petroleum	7290	6570	6630	6440	6310
Chemicals and allied	3730	3640	3940	4010	4190
Metal manufacture	7270	6350	6150	6710	6600
Mechanical engineering	4020	4110	3850	3930	3960
Instrument engineering	1390	490	1370	1430	1560
Electrical engineering	2400	2320	2210	2290	2380
Shipbuilding and marine engineering	6850	6180	6270	5910	5620
Vehicles	3110	3100	3040	3200	3310
Metal goods not elsewhere specified	3870	3800	3610	3740	3700
Textiles	2920	2750	2950	3010	3010
Leather, leather goods and fur	2230	2140	2150	2450	2180
Clothing and footwear	820	810	840	920	950
Bricks, pottery, glass, cement, etc.	5170	4750	4840	4670	4800
Timber, furniture, etc.	3480	3200	3250	3160	3200
Paper, printing and publishing	2480	2270	2450	2500	2610
Other manufacturing	3580	3490	3480	3210	3230

Source: Figures supplied by the Health and Safety Executive

* Figures for the construction industry order are not reported here

undertaken in the remainder of this chapter. There have, however, been a substantial number of studies conducted very largely by industrial psychologists at the plant or firm level. In considering the findings of these studies Kay(15) has suggested that the numerous possible causal variables that they have examined can be usefully grouped under the following sub-headings: the individual who causes the accident; the risks inherent in the work itself; and the working-social environment. It is this framework of analysis that we utilise below in devising individual hypotheses and variables capable of explaining inter-industry variation in accident rates.

The extreme version of the role of the individual in accident causation is the accident prone thesis which holds that certain individuals are _inherently_ more accident prone than others. This view was quite widely held by industrial psychologists in the 1920s and 1930s, but since then has come in for a good deal of criticism.(16) However, a modified version of this argument ('the careless worker' thesis) is still often put forward by employers as one of the prime explanations of industrial accidents; a view that predictably has not been widely accepted by the trade unions. One does not, however, have to accept anything like the full versions of the accident prone or careless worker theses to argue that at least certain personal characteristics of the workforce will be relevant in accident causation. In this regard, for example, it has long been recognised that industrial accidents are very much a male phenomenon, a fact that could be due to systematic differences in the perceived costs of work injuries, in attitudes towards risk bearing, as well as to differences in job experience and occupational and industrial affiliation. Accordingly we anticipate that industries with a relatively high proportion of female workers will have significantly lower accident rates. Although we expect this relationship to be significant it should be clear from the variety of _a priori_ hypotheses outlined above that our results cannot arbitrate between those who advocate the prime importance of worker, as opposed to work, factors in accident causation. This is because the empirical significance of the female worker variable could be accounted for by a work orientated argument (i.e. differences in occupational affiliation) or by a worker orientated argument (i.e. differences in risk bearing attitudes). The nature of our data will not allow us to identify exactly which of these quite different factors underlies our expirical finding; this could only be done if we had disaggregated data cross-classified by industry, occupation, sex, age, etc.(17) The point is that our sub-vectors indicate the basis (i.e. work or worker) on which our variables were computed, rather than the particular a priori argument underlying their potential significance. This is a point to continually bear in mind when considering the implications of our results.

Under this individual worker characteristics sub-heading one also needs to control for the factor of age. In this regard it has long been argued that industrial accidents tend to be disproportionately concentrated among younger aged workers,(18) again possibly because of systematic differences in the perceived costs of work injuries or in attitudes towards risk bearing. However, a number of studies have shown that the frequency of serious accidents is very much less among younger aged workers.(19) The dependent variable in our estimating equations is legally reportable accidents which, as indicated earlier, are more a measure of serious, than of all, industrial accidents. Accordingly, we expect that industries having a relatively high percentage of workers aged 18-24 will have significantly lower accident rates. In addition to age per se it has been argued that accidents tend to be disproportionately concentrated among relatively inexperienced workers, this being a result of their relative lack of familiarity with the nature of the particular work process. For example, a well known study by the now defunct National Institute of Industrial Psychology(20) found that (i) accidents decreased in frequency as the length of service increased; (ii) persons repeating a task which they had done in the recent past had less accidents on it than on tasks which they had done only occasionally; and (iii) on a quickly repetitive task the highest accident rate occurred towards the beginning of the period on the task. Following the approach taken in a number of time series studies of industrial accidents we utilised as a proxy for worker experience the average number of workers engaged per 100 workers employed at the beginning of the period, in the expectation that expanding industries - i.e. those engaging proportionately more workers - would, ceteris paribus, be taking on a greater number of relatively inexperienced workers and hence would have higher accident rates. Finally, we inserted the percentage of unskilled manual employees as a variable in our estimating equation, hypothesising that the well documented, high inter-industry mobility of such workers would give them relatively little experience in any one industry which would make them especially vulnerable to the risk of accidents.

The second sub-vector of potential explanatory variables that we employ here is that for working conditions or working practices that embody a high accident risk component. The basic justification for this sub-vector is provided by the argument that differential accident risks are built into the work process as a result of differences in the amount and nature of the work done. A leading example of this line of argument is a pamphlet by Nichols and Armstrong(21) which contended that accidents are in the main caused by pressure on workers to meet production schedules and deadlines. In other words, their basic contention is that the greater the intensity of the work effort, the higher the accident rate. In an attempt to capture this intensity of work effect we take two variables: the percentage of manual employees under shift work arrangements and the percentage of manual employees working under payment by results schemes. The importance of testing for the influence of these particular variables is evidenced by the number of specific suggestions that they are among the leading work practice(s) causes of accidents.(22) In addition to an 'intensity of work' effect we also need to consider the possible influence of a 'work overload' effect. The role of overtime working has, as we saw in the previous section, been important in explanations of inter-temporal variation in accident rates. This variable would seem equally relevant in an inter-industry context, and so we inserted the percentage of employees working overtime as a variable in our estimating equation.

The third and final sub-vector of variables in our estimating equation was that concerned with the working-social environment. The underlying belief here is that accidents result from the overall nature and quality of the work environment, as well as from pressures on workers emanating from the individual or immediate job task (i.e. the previous sub-vector of variables). The first variable included under this sub-heading is average plant size. This is because it has been quite widely contended that plant size has a direct effect upon job satisfaction and the quality of personal relations at the workplace, with the more regimented working environment of larger plants making for less fulfilling on the job relationships.(23) And one of the specific ways in which this work dissatisfaction may manifest itself is in terms of a higher accident rate. However, if the management of larger plants are relatively alert to such a potential problem then it is likely that they will have taken various forms of action to offset this high accident potential, with the result that the potential does not become a reality. Accordingly, as a result of these potentially offsetting considerations, we make no a priori prediction about the sign on the plant size variable.

Finally under this working-social environment we need to consider the argument that accidents constitute a means of withdrawing from an unacceptable or unsatisfactory working environment, and as such can be interpreted as one particular manifestation of workplace conflict. An early study by Hill and Trist,(24) which reported a positive correlation between accidents and absence, was certainly interpreted in this manner. The findings of the Hill and Trist study, however, have rarely been confirmed in more recent examinations of this issue.(25) Furthermore, the fact that our dependent variable is more a measure of serious accidents would seem to rule out any simple notion that they directly constitute a socially acceptable means of withdrawing from an unsatisfactory working environment. However, it is worth seeing whether the high accident rate industries also score high on other measures of, or proxies for, organised and unorganised conflict. This matter is tested by means of the data set out in Table 2.5.

Spearman (rank) correlation coefficients were computed between industrial accidents and these indices of organised and unorganised industrial conflict. The results were as follows:-

(1) and (2) rs = 0.532 (significant at the .05 level)
(1) and (3) rs = 0.505 (significant at the .10 level)
(1) and (4) rs = 0.594 (significant at the .05 level)
(1) and (5) rs = 0.183

The results obtained indicate that the high accident rate industries, in which there are a disproportionate number of unionised employees (see Table 1.2) have significantly high levels of strike activity, certificated sickness absence and voluntary absence. In short, the high accident rate industries also have a disproportionate amount of organised (strikes) and unorganised (voluntary and certificated sickness absence) industrial conflict. The implications of these findings for the operation of the safety representative and safety committee functions will be discussed in the final section of this chapter. For the moment, however, we simply take strikes as the most obvious and dramatic example of industrial conflict and enter as a variable in our estimating equations the number of industrial stoppages per 100,000 employees. Although this variable is hypothesised to bear a positive relationship to industrial accidents it should be seen not as embodying a direct causal relationship, but rather as proxying the more general conflict potential of the working-social environment.

This completes our list of potential explanatory variables, but before proceeding to examine the results obtained a brief word is necessary about the absence of a wage variable from our estimating equation. The average wage for the industry could be argued to represent the human capital embodied in the average worker of that industry and hence one could hypothesise that the higher the wage the greater the effort that would be devoted to protecting the human capital it represents. However, the accident rate, as an indicator of the nonpecuniary disbenefits of the job, is potentially one of the determinants of the wage in that a higher risk of accidents should, *ceteris paribus*, induce a higher wage to compensate for that risk; indeed some evidence to this effect will be cited in the next chapter. In short, the fact that the wage rate is not determined independently of the accident rate requires the use of a simultaneous equation procedure. The concern here is solely with attempting to isolate the *exogenous* influences on the accident rate of each industry and hence there is no place for a wage variable.

An ordinary least squares regression was estimated on the accident rate per 1,000 employees for 113 MLH or 3 digit industry groups in the manufacturing sector for the year 1970. The results are set out in Table 2.6.

The results of this exercise were quite encouraging; the overall level of explanatory power achieved by the equation was extremely respectable for a cross section study, virtually all the variables entered with the hypothesised sign and five of them attained some level of statistical significance. Turning to the results for the individual variables, the female employees variable

Table 2.5: Industrial Accidents, Organised and Unorganised Conflict Variables, 1970*

Industry	Industrial Accidents (1)	Strike Incidence per 100,000 Employees (2)	Voluntary Absence (3)	Certificated Sickness Absence (4)	Late Arrival or Early Finish (5)
Food, drink and tobacco	38.2	16.6	2.6	4.3	2.2
Coal and petroleum products	54.8	20.5	2.4	3.6	3.6
Chemicals and allied industries	35.6	17.8	1.8	4.6	2.4
Metal manufacture	78.9	55.1	6.1	6.4	8.2
Mechanical engineering	38.6	42.9	3.8	4.5	7.8
Instrument engineering	15.1	23.8	1.7	3.1	6.5
Electrical engineering	24.6	31.5	3.8	3.5	6.1
Shipbuilding and marine engineering	72.4	60.8	7.7	5.9	8.8
Vehicles	34.5	53.5	4.4	5.3	8.1
Metal goods not elsewhere specified	39.7	27.7	4.3	3.4	7.2
Textiles	26.8	13.4	3.6	3.9	3.6
Clothing and footwear	7.9	5.4	1.7	2.3	7.2
Bricks, pottery, glass and cement	54.2	23.5	3.2	4.2	4.7
Timber and furniture	31.3	17.7	3.0	4.1	5.7
Paper, printing and publishing	24.8	11.5	2.7	4.0	5.6

Other manufacturing industries	34.6	25.4	4.5	5.1	6.6

Source: R Bean, "The Relationship Between Strikes and Unorganised Conflict in Manufacturing Industries", *British Journal of Industrial Relations*, March 1975, p.100; *Annual Report of the Factory Inspector*, 1970.

* There were no observations reported for the Leather, leather goods and fur industry order.

entered with the anticipated negative sign and was statistically significant. The age variable was also significant and negatively signed indicating that the higher the percentage of younger aged (18-24) workers in an industry the lower the accident rate. The direction of this particular relationship was undoubtedly due to the fact that our dependent variable was a measure of serious, rather than of all, accidents. The engagements variable entered with a positive sign indicating that industries with expanding work-forces had higher accident rates, but this result was nowhere near being statistically significant. The most highly significant variable in the equation was that for skill, with the higher the percentage of unskilled workers in an industry the higher the accident rate.

The payment by results variable was positively signed and statistically significant indicating that the higher the percentage of manual employees working under such arrangements the higher the accident rate. The shift work variable was also positively signed indicating that the higher the percentage of manual employees on shift work the higher the accident rate. This result was not, however, statistically significant which may have been due to something of a multi-collinearity problem (the simple r between payment by results and shift work was 0.61) and/or our inability to differentiate between the different types of shift work systems, some of which may embody far greater accident risk potential than others. Although the extent of overtime working proved important in our earlier examination of intertemporal change in the accident rate it performed 'perversely' here as a proxy for a work overload effect. The variable entered with a negative sign suggesting that the higher the proportion of manual employees working overtime the lower the accident rate, although the coefficient was not statistically significant. This unexpected result may have been due to the form in which the variable was entered in the equation, i.e. as the percentage of manual employees working overtime. The total number of overtime hours worked in any given industry over a given period of time is a function of the number of employees working overtime and the number of overtime hours worked by each individual. And if a higher number of people working overtime reduces the average number of overtime hours worked by an individual, and it is the <u>hours</u> dimension that is most relevant in accident causation, then the above result would suggest the need to respecify the

Table 2.6: Regression on Accidents per 1,000 Employees at MLH Level for Manufacturing in Britain, 1970

(Standard errors in parentheses)

Variables		
Constant	11.86993	
Female	- 0.2547592	**
	(0.11572)	
Age	- 1.214151	**
	(0.58165)	
Engagements	0.1134810	
	(1.87268)	
Skill	2.173464	***
	(0.33469)	
Payment by Results	0.2260553	**
	(0.10960)	
Shift Work	0.1721166	
	(0.12175)	
Overtime	- 0.1029275	
	(0.06716)	
Plant Size	- 0.0096541	
	(0.01110)	
Strikes	0.1453436	*
	(0.08531)	
R^2	0.63217	
F	19.66908	
N	113	

* significant at the .10 level (2 tail test)
** significant at the .05 level (2 tail test)
*** significant at the .001 level (2 tail test)

the overtime variable in an hours form.

The plant size variable entered with a negative sign suggesting that there were rather less accidents in larger sized plants, but this result was nowhere near being statistically significant. The weakness of the plant size effect here may simply be due to the offsetting effect of the two opposing hypotheses that we outlined in our *a priori* specification. Alternatively, its statistical weakness may have resulted from the form in which the variable was specified. The variable was entered as the mean number of employees per establishment which may not be the appropriate functional form if there exists any significant discontinuity in the accident-size relationship above or below a particular size band. Finally, the strikes variable entered with the predicted positive sign indicating that the greater the number of strikes the higher the industry accident rate. This relationship was statistically significant, although only at the rather lowly .10 level.

The above results seem sufficiently encouraging to suggest the value of elaborating and extending the basic model presented here in an attempt to more fully identify the factors associated with high accident rates. The possible extensions that could usefully be made include changes in the form of a number of the existing variables (e.g. overtime, plant size), together with the inclusion of some extra variables under the three sub-vectors outlined. However, the most potentially valuable extension to the model would seem to be the addition of a sub-vector of possible *preventive* measures. Such preventive variables could be *external* in nature such as the inspection rate of the Factory Inspectorate, or *internal* ones designed to measure or proxy the extent of management commitment to the safety function. In practice such measures or proxies will be difficult to devise and interpret, for reasons that will be discussed in later chapters, but the potentially high pay-off to such an exercise would certainly seem to warrant a serious attempt being made in this regard.

Finally, in devising and testing a model concerned with the factors that account for variations in accidents across a broad range of industries one should not lose sight of the potential importance of certain *industry specific* factors in accident causation. There are a number of industries where it has long been held that the specific nature of the production process, working practices, etc., make *inevitably* for above average accident rates. The coalmining industry is an obvious example in this regard. The inherent, high accident risks of this industry account for the fact that special health and safety legislation, often providing for union-management arrangements well in advance of the rest of industry, has long been in operation in the coalmining industry in Britain and other countries. The construction industry has also been singled out as a special case in the matter of industrial accidents. In Britain there has been a great deal of discussion about the adverse implications of labour only sub-contracting arrangements ('the lump') for health and safety in the construction industry. Although there is little conclusive evidence to indicate that this form of working arrangement directly leads to high accidents there has been a great deal of discussion about the extent to which statutory safety requirements effectively apply to these self employed persons.(26) Certainly the Phelps Brown Committee of Inquiry stated that "HM Factory Inspectorate report considerable difficulty in applying safety legislation to the self-employed and find this tends to bring the regulations into disrepute."(27) As a final example of the potential importance of industry specific factors one can point to the offshore oil industry. Indeed, as Lock and Smith have stated, "measured by the rates per man of injuries and death, the most dangerous occupation in Britain at present is work off-shore in the oil exploration industry. Oil rig workers have an annual fatality rate of 2-3 per 1,000 - ten times the rate for coalmining and fifty times that in factories in general."(28) The following figures, which were supplied by the Department of Energy, reveal the extent of the accident problem in the off-shore oil industry:-

Year	Deaths	Serious Accidents
1971	4	17
1972	3	17
1973	3	22
1974	12	25
1975	10	75
1976	17	57
1977 (to end of July)	8	12

There has been special safety legislation covering off-shore activities,(29) but in the light of these figures an order was laid before Parliament on August 3, 1977, which meant that the standards of occupational health and safety protection equivalent to those on-shore would be provided for those working on and in connection with off-shore installations. The result of this order was that from September 1977 workers in the off-shore gas and oil industry were covered by the general duties provisions of the Health and Safety at Work Act 1974. However, the safety representative/committee regulations have not been extended to these off-shore installations, a point that was critically noted in a recent report which compared the UK and Norwegian systems of regulation of health and safety off-shore. (30)

The Implications of the Inter-Industry Analysis

As the major substantive part of this chapter has been concerned with an analysis of inter-industry variation in accident rates it is therefore appropriate to conclude here by looking at the implications of this analysis for the operation of the new safety representative and safety committee arrangements. The first point to be considered in this regard concerns our finding that the industries where there is most scope and need for improved health and safety performance, at least in the sense of achieving a lowering of accident rates, are those that are highly organised (see Table 1.2) and have high levels of organised and unorganised conflict (see Table 2.5). It needs to be asked whether these findings have certain pessimistic implications for the likely operating effectiveness of the safety representative and safety committee functions in bringing about improvements in workplace health and safety.

In considering the above question it is important to note that health and safety is typically held to be a subject area in which a joint problem solving approach by unions and management should yield benefits to both sides. In Britain this is the rationale for health and safety being considered a subject suitable for 'consultation' rather than 'negotiation', a matter that we discuss at some length in Chapter 5. The necessary prerequisites for successful joint problem solving are held to include attitudes of mutual trust, friendliness and respect between the relevant union and management representatives and a supportive industrial relations climate in the sense of an overall, co-operative union-management relationship.(31) The question here is whether the high levels of conflict identified above necessarily mean the absence of the mutual trust, friendliness and overall co-operative relationship between unions and management that is so essential for successful joint problem solving? In other words, are the high accident rate industries, where there is the most obvious need for an improvement in workplace health and safety, the very industries where this is least likely to come about due to the lack of the co-operative union-management relationships that will facilitate the successful introduction and operation of the safety representative and safety committee functions?

In seeking to answer these questions there are a number of points that we would wish to make. First, that within the industries with high levels of industrial conflict we are likely to find this conflict disproportionately concentrated in a relatively small number of employment establishments. This point was well illustrated by a recent Department of Employment study of strike activity. This study, which examined data for the high strike years 1971-3, found that only 5 per cent of manufacturing plants had experienced a stoppage in those years, and that of these over 66 per cent had experienced only one stoppage while "a very small minority of plants had a large number of stoppages."(32) The importance of this sort of finding, that within industries conflict is disproportionately concentrated in a relatively small number of workplaces, is that it is the nature of individual workplace, rather than industry level, union-management relationships that is important in fostering successful joint problem solving arrangements and behaviour. Furthermore, the presence of industrial conflict at the level of the individual workplace does not necessarily mean the existence of 'poor' or 'uncooperative' union-management relationships. This point is evident, for example, from the responses of our sample of safety representatives which are reported in Chapter 5.

On the other hand, one fact that is likely to constitute an important constraint on the ability of safety representatives and safety committees to bring about improvements in workplace health and safety is the very marked degree of stability of relative industry rankings in the accident league through time, i.e. our finding that the high(low) accident rate industries in any one year are very much the high(low) accident rate industries in other years. The strength of relative industry rankings in the accident league table across the period 1974-78, for example, indicates that there are very strong forces making for high accident rates that, at least on an industry by industry basis, are almost invariant with respect to time. The sorts of forces that are relevant here include those identified in our inter-industry regression estimates. The safety representatives and safety committees will have to confront these sorts of forces and their obvious strength suggests that it is extremely unlikely that the safety representatives and safety committees will be able to bring about improvements in workplace health and safety that will be reflected in changes in the relative ranking of industries in the accident league table, i.e. one is unlikely to see the traditionally high accident rate industries significantly shift down the accident league table.

The above conclusion should not cause the reader to be unduly pessimistic about the potential impact and effectiveness of the safety representative and safety committee functions in improving workplace health and safety. In our view the successful operation of these functions is still, at least in principle, capable of producing the following changes:

(i) An absolute, if not relative, fall in the reported accident figures for the high (and lesser) accident rate industries. If such an effect does come about it is likely to be disproportionately concentrated within a particular sub-category (by cause) of reported accidents. The sub-category (by cause) of reported accidents that the safety representatives/committees have the most potential for influencing is a matter that warrants serious examination.

(ii) The legally reportable accidents (i.e. that involve a fatality or absence from work for more than 3 days) that we have used in the analysis of this chapter constitute, according to earlier reported Department of Employment estimates, a very small proportion of all industrial accidents that occur. A reduction in these more numerous, if less individually serious, non-reportable accidents may well result from the successful operation of the safety representative and safety committee functions.

(iii) Industrial accidents are not the sole, or even in the view of some commentators the most important, measure of the state of workplace health and safety. Accordingly, there may be other safety (and health) indicators that will usefully reveal the operational effectiveness (or lack of) of the safety representative and safety committee functions. In this regard one might usefully view the health and safety performance of an individual employment establishment or industry as being an outcome of the following interaction:-

(1) health and safety <u>record</u>	x	(2) health and safety <u>practices</u>
(a) Accidents of varying severity (and near misses)		(a) Extent to which organisations produce and comply with their own 'codes of practice' in relation to
(b) Diseases		

(c)	Extent of compliance with legal requirements to eliminate or minimise hazards	potential hazards. These best practice standards, over and above the statutory minima, can be observed in the nature of rules and procedures, work systems, and training and monitoring methods in relation to health and safety matters.

The importance of using this multiplicative term, rather than simply 1(a) above, to review the health and safety performance of an establishment or industry arguably follows from the fact that such a substantial proportion of all accidents that occur are not in fact preventable by any reasonably practicable precautions that could be taken by management and/or employees. In this regard a Factory Inspectorate study in Britain in 1968, for example, suggested that the number of accidents in this non-preventable category could be as high as 50 per cent. (33) Accordingly, if only half of the reportable accidents are due to both breaches of the law (only 18 per cent according to the Factory Inspectorate study) and the absence of reasonably practicable management-employee precautions then it could be argued that health and safety practices are a more reasonable, reliable and useful indicator of the quality of the overall health and safety performance of an establishment or industry than the accident component of the establishment or industry's health and safety record.(34) This is because the former is held to be much more under the control of the establishment or industry and is therefore a more accurate indicator of the initiatives that have been taken in order to try to improve the overall health and safety performance of the establishment or industry. This distinction between the health and safety record and health and safety practices of an establishment or industry certainly has, at least on the basis of the results of one recent study in the United States,(35) important implications for judging the effectiveness of experiments to improve the quality of working life. This is a matter that could be usefully followed up in future research in Britain.

(iv) Finally, studies of 'successful' joint problem solving arrangements and behaviour at the workplace have found that participants tend to define 'success' less in terms of solving specific, individual issues and more in terms of bringing about improved inter-personal and inter-party (unions and management) relationships;(36) the latter effect having favourable implications for the larger, industrial relations position at the individual workplace. This is a line of influence that is considered in more detail in Chapter 6.

These various hoped for aims and changes may be less dramatic than bringing about a significant change in the relative ranking of industries in the reported accident statistics, but they would seem, at least in our view, rather more realistic and are certainly far from being unimportant effects. Accordingly, it is to such possible effects that future researchers could usefully address themselves.

NOTES

1. Olga Aikin and Judith Reid, Employment, Welfare and Safety at Work, Penguin, Harmondsworth, 1971, p. 365.
2. Roy Lewis and Geoff Latta, "Compensation for Industrial Injury and Disease", Journal of Social Policy, Vol. 4, No. 1, January 1975, p. 27.
3. Report of the Safety and Health at Work Committee (Robens), 1970-72, Cmnd 5034, HMSO, London, 1972, p. 139.
4. Cmnd, op. cit., p. 141.
5. See, for example, G Calabresi, The Costs of Accidents: A Legal and Economic Analysis, New Haven, 1970.
6. Evidence of the Department of Employment to the Health and Safety at Work Committee (Robens), 1970-72, Selected Written Evidence, Vol. 2, HMSO, London, 1972, p. 290.
7. Evidence of the Department of Employment, loc. cit.
8. E H Phelps Brown, The Growth of British Industrial Relations, Macmillan, London, 1959, p. 72-3.
9. Robert S Smith, "Intertemporal Changes in Work Injury Rates", Proceedings of the Industrial Relations Research Association, Winter 1973.
10. Smith, op. cit., p. 172.
11. G R Steele, "Industrial Accidents, An Economic Interpretation", Applied Economics, Vol. 6, 1974.

12. James R Chelius, Workplace Safety and Health: The Role of Workers Compensation, American Enterprise Institute for Public Policy Research, Washington D C, 1977, p. 14.

13. See, for example, Walter Oi, "On the Economics of Industrial Safety", Law and Contemporary Problems, 1973, p. 683.

14. Although see, for example, Chelius, op. cit. p. 81-83.

15. H Kay, "Accidents: Some Facts and Theories" in Peter B Warr (ed.) Psychology at Work, Second Edition, Penguin, Harmondsworth, 1978, p. 104.

16. See, for example, A B Cherns, "Accidents at Work", in A T Welford et. al. (eds.), Society: Problems and Methods of Study, Routledge and Kegan Paul, London 1962, p. 248-53.

17. For a study using such data see Alan E Dillingham, "Sex Differences in Labor Market Injury Risk", Industrial Relations, Vol. 20, No. 1 Winter 1981.

18. J Surry, Industrial Accident Research, University of Toronto Press, Toronto, 1968, p. 12-13.

19. Surry, op. cit., p. 13-14.

20. National Institute of Industrial Psychology, 2,000 Accidents, London, 1971.

21. Theo Nichols and Peter Armstrong, Safety or Profit: Industrial Accidents and the Conventional Wisdom, Falling Wall Press, Bristol, 1973.

22. See, for example, ILO, Payment by Results, Geneva, 1951, p. 151.

23. See, for example, Geoffry K Ingham, Size of Industrial Organisation and Worker Behaviour, Cambridge University Press, London, 1970, p. 17.

24. J M M Hill and E L Trist, "A Consideration of Accidents as a Means of Withdrawal from the Work Situation", Human Relations, Vol. 6, 1953.

25. See, for example, P B Beaumont, "The Relationship Between Industrial Accidents and Absenteeism", Industrial Relations Journal, Vol. 10, No. 3, Autumn 1979.

26. See, for example, B I Mordsley, "Some Problems of the Lump", Modern Law Review, Vol. 38, September 1975.

27. Report of the Committee of Inquiry Under Professor E H Phelps Brown into Certain Matters Concerning Labour in Building and Civil Engineering, Cmnd 3714, HMSO, 1968, parag. 370.

28. Stephen Lock and Tony Smith, The Medical Risks of Life, Penguin, Harmondsworth, 1976, p. 148.

29. Jonathan Kitchen, Labour Law and Offshore Oil, Croom Helm, London, 1977, Chapters 5 and 6.

30. Richard Howells and Brenda Barrett, "Occupational Health and Safety in the North Sea", New Law Journal, April 2, 1981, p. 354-6.

31. Richard E Walton and Robert B McKersie, A Behavioural Theory of Labor Negotiations, McGraw Hill, New York, 1965.

32. C T B Smith, et. al., Strikes in Britain, DE Manpower Paper No. 15, 1978, p. 54.

33. See Department of Employment Written Evidence, Loc. cit.

34. See Success and Failure in Accident Prevention, Accident Prevention Advisory Unit, Health and Safety Executive, 1976.

35. Paul S Goodman, Assessing Organisational Change: The Rushton Quality of Work Experiment, Wiley and Sons, New York, 1979, Chapter 12.

36. Richard B Peterson and Lane Tracy, "Testing a Behavioural Theory Model of Labor Negotiations", Industrial Relations, Vol. 16, No. 1, February 1977.

Chapter Three

UNION INVOLVEMENT IN WORKPLACE HEALTH AND SAFETY PRIOR TO THE 1974 ACT

In the previous chapter we examined various dimensions of the long-standing problem of industrial accidents in Britain. This chapter follows up that discussion by considering the nature of the traditional forms of union response to the presence of this problem. The basic perspective adopted here is that, in contrast to what has often been argued or implied in the trade union literature,(1) there has been no clearly formulated, comprehensive union policy on the matter of accident prevention. The absence of such a policy is a matter strongly emphasised by Williams in the following terms,(2)

> There is a large amount of activity, a great proportion of which is effective, on specific matters relating to safety and health. Decisions have been taken, at various times, on particular issues such as the extension of statutory standards to fields of employment not covered; the strengthening of the Factory Inspectorate; the prescription of a particular complaint as an industrial disease; the need for safety committees and other specific issues. All these policy decisions on specific questions are designed to reduce accidents and ill-health at work. They are an important contribution to be expected from trade union activity, they are the result of constant pressure and consideration by trade union and TUC officials. If these policy decisions on specific questions were to be set down and arranged suitably they would amount to an accident prevention programme of a kind. But they would not have been formulated in relation to each other and integrated into an overall plan which is a fundamental part of an effective policy.

Williams then went on to provide an historical over-view of union activities in relation to accident prevention, and workplace health and safety matters more generally. In this exercise he explicitly considered three periods of time - 1800 to 1860; 1860 to 1900; 1900 to 1960 - and stressed that over time "the basic attack of the movement had ... developed into four avenues - improvements in the accident compensation system; improvements of existing statutory standards; extension of statutory standards to new fields; and exploitation of joint machinery with employers and Government departments for these purposes."(3) In considering the nature of union responses to the problem of industrial accidents, as well as other indicators or manifestations of unsatisfactory workplace health and safety, we utilise a somewhat more detailed schema than that of Williams, namely that which was presented in the introductory chapter. This schema, in a slightly re-arranged order from that presented earlier, is as follows:-

(i) Lobbying and supporting Government legislation to regulate workplace health and safety conditions.

(ii) Aiding and representing workers in accident compensation claims.

(iii) Negotiating compensatory wage differentials for workers in high risk jobs and industries.

(iv) Negotiating safety provisions of a preventive and/or compensatory nature in collective agreements.

(v) Pushing for the establishment of joint union-management health and safety committees.

The general line of argument to be presented here is that there was union activity under all of these headings in Britain prior to the passage of the 1974 Act, but that it was overwhelmingly concentrated under the first three headings. As well as illustrating this general proposition a major portion of this chapter is devoted to developing and testing a model designed to identify the relevant characteristics of plants that had voluntarily established joint health and safety committees prior to the 1974 Act. This exercise, which is the major substantive contribution of the chapter, is important in its own right as well as being important in providing a number of insights that will be drawn upon in Chapter 6 where we consider the position of joint health and safety committees since the passage of the 1974 Act.

In using the above schema to organise the

material presented in this chapter we should stress again that the unions have not had a clearly formulated, inter-related, comprehensive policy in relation to accident prevention or workplace health and safety more generally, and hence the reader should not draw any 'ordered union approach' connotations from the use of the schema. Furthermore, although we believe that this schema encompasses all major union channels of influence that have been directly concerned with this subject area there are other union demands and activities that may indirectly have had implications for the state of workplace health and safety. Union demands for shorter working hours could certainly be seen in this light. More importantly, the above schema could perhaps be expanded to include the Webbs' trade union method of 'mutual insurance'. This function, which covers the benevolent or friendly society side of trade unionism, is often held to be of relatively little contemporary importance, largely as a result of the increased Government provision of welfare benefits. However, a study by Latta and Lewis(4) found that fully 33 of the 41 unions they examined provided accident and sickness benefits for their members. Moreover, figures for the year 1970 revealed that registered unions paid out £3,885,000 to their members in sickness and accident benefits, a not insignificant sized supplement to state provided benefits in this area. These figures, together with the fact that a number of unions such as the General and Municipal Workers Union, only started to pay such benefits during the 1960s suggest that the 'mutual insurance' method has not declined to anything like the extent that is often supposed.

As a final, introductory point it is important to note that there are adjustment mechanisms or responses to the presence of high accident rates, even in the absence of trade union organisation. The obvious candidate in this regard is quits or turnover - i.e. there will be above average quit rates in plants and industries having above average accident rates. There is certainly evidence of such an adaptive response on the part of individual employees,(5) but the general perspective adopted here is that this 'exit' mechanism becomes of relatively less importance with the introduction of the 'institution of voice' in the labour market provided by trade unionism and collective bargaining.(6)

Lobbying for Health and Safety Legislation

Trade unions in Britain and elsewhere have long

sought to influence the passage of legislation designed to assist the position of their members. In Britain there is considerable evidence of such union activity in relation to health and safety legislation. The long standing nature of this activity is well evidenced by the Webbs detailed description of the attempt of individual unions to influence public and parliamentary opinion in favour of general and industry specific health and safety regulations during the 19th and early 20th century. According to the Webbs,(7)

> The long and elaborate code of law which now governs employment in the factory and workshop, the bakehouse and printing office, on sea and in the depths of the mine, is itself largely made up of the Common Rules designed for the protection of the operatives health, life or comfort, which have been pressed for by Trade Unions, and have successively commended themselves to the wisdom of Parliament ... we attribute the rapid development of this side of Trade Unionism to the discovery by Trade Union leaders that it is the line of least resistance. Middle class public opinion, which fails as yet to comprehend the Common Rule of the Standard Rate and is strongly prejudiced against the fixing of a Normal Day, cordially approves any proposal for preventing accidents or improving the sanitation of workplaces.

The Webbs' view was that unions in coalmining and cotton textiles were particularly active and successful in lobbying for the passage of health and safety legislation during the time period which they examined. In more recent times individual unions still retain sponsored M.P.s who will concern themselves with such matters, but the major publicity and lobbying functions on behalf of health and safety regulation have tended to pass from individual unions to the TUC. In this regard the general council of the TUC came to favour statutory based joint health and safety committees from the mid-1960s and gave evidence to the Robens Committee to this effect. More recently it gave evidence to the Pearson Committee of inquiry into compensation for personal injury, urging the need for changes and reforms to the Industrial Injuries Scheme, the system of common law damages and the coverage of employer sick pay schemes.(8) In addition to presenting such evidence on behalf of the trade union movement the TUC has also sought to influence the more day to day administration of health and safety regulations. The frequent demands for an increase in the size of the factory inspectorate being an obvious example in this regard. The extent of union success in influencing the nature of changes in health and safety regulations is a difficult matter to gauge. A number of general assessments of the TUC pressure group function have suggested that it has not been a particularly powerful force in influencing matters of policy determination.(9) There is certainly no obvious evidence to suggest that their influence in the specific area of health and safety has been above that of their average performance. For example, the CBI certainly appeared to be more satisfied with the recommendations of the Robens Committee than was the TUC; the latter were certainly highly critical of quite a number of omissions in the report, not the least of which was the absence of a recommendation for a statutory based system of joint health and safety committees.(10) More recently, however, trade union influence in both policy determination and administration in Britain has been seen to be substantially enhanced by participation in tripartite bodies with statutory responsibilities for policy implementation; the Health and Safety Commission established under the terms of the Health and Safety at Work Act 1974 being just such a body. However, the Health and Safety Commission has been particularly adversely affected by the public expenditure cuts of the present Government, a fact which raises obvious questions about its 'political' strength in the future. This is a matter that we return to in our concluding chapter.

Aiding and Representing Workers in Accident Compensation Claims

The rule books of most unions empower them to provide legal assistance to their members in connection with employment problems. This is an important but much overlooked union function in that,(11)

> It is not in the mass campaign, or the lock-out that the worker is most apt to feel isolated and unbefriended, for in these he can count on the moral and financial support of his fellows. It is rather in the personal disaster which visits him alone. Even when there is statutory provision for helping him, the impersonal complex apparatus of the modern state through

which he has to pursue it may seem chilling and deterrent The reassurance and practical help now being provided by the Union was a necessary adjustment to the extended operations of the state itself, and probably only the member who has known the sudden collapse of his sense of security which follows a personal disaster knows just how high to rate the ready information and advice which lies to hand.

In Britain in the post war years an injured employee has been able to claim damages at common law, as well as receiving benefits under the terms of the Industrial Injuries Scheme. Social insurance benefits are payable under the Industrial Injuries Scheme if an employee suffers either an industrial accident, or a prescribed industrial disease. During absence from work a flat rate injury benefit may be paid weekly, together with allowances for dependents and an earnings related supplement. If the injured person suffers any loss of physical or mental faculty he may become eligible for disablement benefit. There is also provision for a death benefit payable to the dependents of any person killed in an industrial accident. The total expenditure on benefits from the Industrial Injuries fund in 1972, for example, was £135 million. A detailed study by Latta and Lewis(12) estimated that unions support approximately 10,000 appeal cases per year under the Industrial Injuries Scheme. In addition to the benefit provisions of the Industrial Injuries Scheme, an injured employee may also claim damages under the common law, alleging that his injury resulted from negligence or breach of statutory duty on the part of his employer. Employers are required by law to insure themselves against such common law claims and hence the negotiations over them usually take place with insurance companies, rather than with the employers directly concerned. Some figures cited by Latta and Lewis(13) suggest that in 1971 unions negotiated about 50,000 successful common law claims for their members, which produced damages totalling around £20 million. Although Latta and Lewis pointed out that unions only gain common law damages in approximately 10 per cent of cases involving their members they suggested that the probability of a successful claim for an employee was likely to be very much less in the absence of union representation. Certainly the cost of union legal services has been rising faster than any other heading of union expenditure in recent times, and the vast bulk of the legal workload of most unions is provided by industrial accident compensation claims.

The existing extent of union involvement in accident compensation claims should be seen in the larger context of union criticisms of the present day arrangements for accident compensation. There were resolutions calling for change to the system of compensation passed at the annual meeting of the TUC in 1966 and 1972 and, as we noted earlier, the TUC gave evidence to this effect to the Pearson Committee of inquiry. However, the recommendations of the Pearson Committee were for improving, rather than replacing, the present system of arrangements. (14) These recommendations were greeted with a not inconsiderable amount of union criticism, but are likely to ensure that unions continue to play an important role in representing members in compensation claims. In view of this likelihood it is important to note Latta and Lewis's criticism that "... too few unions systematically use their legal cases to pinpoint where accidents occur, or to assess the incidence or types of accident. Moreover few use their legal cases as a basis for collective bargaining."(15) In other words, there appears to have been relatively little positive spillover from this quite extensive union involvement in the compensatory side of health and safety to the preventative side. This criticism appears to have been increasingly accepted by the unions themselves from the mid-1960s,(16) so that it is to be hoped that such a development will increasingly take place as the potential, preventative role of the safety representative and safety committee provisions of the 1974 Act is increasingly drawn upon.

Negotiating Compensatory Wage Differentials

The theory of compensatory wage differentials suggests that jobs with disagreeable characteristics will command higher wages, *ceteris paribus*, because "the whole of the advantages and disadvantages of the different employments of labour and stock must, in the same neighbourhood, be either perfectly equal or continually tending toward equality."(17) Specifically the higher the accident risk of a job, the higher the wage that must be paid to compensate workers for bearing that risk. Alternatively, the employer can invest more in reducing the risks associated with the job and thereby reduce the need for (or size of) the compensating differential; indeed this is the desired, indirect effect of such differentials. What is the available evidence regarding the existence and size of such wage differentials? A recent review article (18) listed seven different studies of the size of compensatory wage differentials for dangerous jobs. Although only one of these studies was for Britain, (19) its findings were essentially similar to the others so that we report here the major conclusions of the group of studies as a whole. First, virtually all the studies which used a 'risk of death' variable found it to have a positive and statistically significant coefficient in the wage regression, but the results for the risk of non-fatal injury were much less clear cut. Secondly, the fatal injury related wage differentials paid to the average worker in each study were estimated to be some 1-4 per cent of earnings. However, the level of risk faced by the workers in these studies varied considerably so that this narrow band of average differentials seems to imply a wide variation in the total wage compensation associated with an additional occupational death. Furthermore, studies using fatal accident rates by industry estimated much larger wage-risk trade offs than studies using fatal accident risks by occupation. Subsequent to this review article another study in the United States reported an important interaction between fatal accident risk and union status.(20) That is, compared to non-union workers, union members received substantially higher fatal accident premiums.

The existence of such compensating wage differentials does not necessarily mean that their size adequately reflects the extent of on the job risk. Indeed the possibility of fully compensating differentials is frequently held to be unlikely on the grounds that wage earners as a body tend to systematically under-estimate the disadvantages of risky jobs. This is because "until a man has had experience of a certain kind of work, he is unlikely to know that it is dangerous, and then the damage is often done. And even when the danger is known, most people are too inclined to suppose that they can escape damages which overcome others."(21) However, the theory of compensating wage differentials does not require <u>all</u> workers to correctly perceive the riskiness of a job, rather the market adjustment mechanism only requires the <u>marginal</u> worker to have the correct perception of risk. In view of this apparently 'lesser' requirement one could envisage unions as playing an important information gathering role in estimating the risk-wage trade-off for the marginal worker. However, Zeckhauser(22) has suggested that unions may be unwilling or unable to perform such a role in that union leaders may not wish to have wage premiums identified as 'risk pay' since it would give the appearance of condoning risk to their members. Furthermore, trade union leaders do not personally bear the full cost of inadequate risk perception, so that the incentive to properly estimate the risk of work injuries is lessened. It has also been reported that the size of union negotiated risk differentials in the United States tends to change very little through time, which suggests that such compensation is slow to adapt to increasing knowledge and information about the extent of job hazards.(23) It is in fact this belief that risk premiums do not fully compensate workers for the extent of on the job hazards, and as such do not provide sufficient incentive for employers to invest in safety prevention measures, that provides the basic rationale for Government intervention in the area of workplace health and safety. However, the above *a priori* reservations about the union role in this regard would seem to have been called somewhat into question by recent estimates of the union/non-union wage differential in the United States which emphasise the role of poor quality working conditions in accounting for much of the differential.(24)

These risk compensation premiums can be built into the formal wage structure through means of job evaluation, or can result from a more informal process, such as that of workplace level bargaining between shop stewards and supervisors. However, before we discuss this workplace level bargaining in the next section it is important to note that a number of unions in Britain have recently issued policy statements discouraging the negotiation of such premiums.(25) This is suggestive of a significant change in union attitudes since the passage of the 1974 Act, as the TUC guide to negotiators in 1971 explicitly stated that collective agreements should contain extra payments for work in difficult, dangerous or unpleasant conditions.(26)

Negotiating Safety Provisions in Collective Agreements

There appears to have been relatively little *formal* collective bargaining over health and safety matters in Britain. This point was made in a paper by Lewis who suggested that it had been very much confined to a few special cases, namely the Fire Brigades Union, the Post Office Engineers and a number of plant or company level agreements negotiated with large American multi-nationals.(27) There are a number of possible reasons that may be cited to explain the relative lack of formal collective bargaining over health and safety matters in Britain. The first reason often given is that workers are relatively unwilling to strike over health and safety issues, possibly because of the high variance in the exposure of different workers in the same plant to unsafe or risky jobs and conditions. There certainly appear to have been few strikes directly over health and safety issues in Britain in recent times; over the period 1966-74, for example, strikes over 'working conditions' issues (of which health and safety matters would only be one part) ranged from only 5 per cent in 1969 to 8.2 per cent in 1973.(28) This contention should, however, be seen against the background of the Webbs' claim that, during the late 19th century, "... there is no subject on which workmen of all shades of opinion, and all varieties of occupation, are so unanimous, and so ready to take combined action, as the prevention of accidents and provision of healthy workplaces."(29) A second reason that is often put forward to explain the absence of such bargaining is that the strong similarity of union-management aims and interests in relation to health and safety matters makes it a subject suitable for joint consultation rather than collective bargaining. This is a belief that we explore in some detail in Chapter 5. A third reason is that the legal environment prior to the 1974 Act strongly constrained the opportunities for such union-management interaction, which is a matter that we examine in the following chapter. However, the particular reason we wish to emphasise here has to do with the nature of bargaining structure. The fact that there exists a close relationship between the level at which bargaining is conducted and the subject matter of that bargaining is well known. According to Weber,(30)

> Some issues, like wages, have market-wide implications and must be handled within expanded bargaining structures to insure their effective resolution. Other issues, such as pensions and insurance plans, are best treated on a company-wide basis because of the need for uniformity created by actuarial and administrative considerations. Finally, there are questions of work rules, safety, wash-up time and other minutiae of industrial relations that are essentially

local in nature and must be related to the conditions that prevail in a particular plant or department.

This perspective led Weber to argue that the particular subject matter of collective bargaining was an important determinant of bargaining structure. However, we would tend to argue that the line of causation or influence can also run the other way, with a well established structure of collective bargaining very much shaping the scope of collective bargaining. Certainly the Webbs'(31) discussion indicated that health and safety matters were frequently covered by collective bargaining during the years of the late 19th century, a period of predominantly district or local area based bargaining in Britain. However, as a basis of the system of collective bargaining in Britain became increasingly a multi-employer, industry based one this had the effect of limiting the scope or range of such bargains to essentially wages-hours matters.(32) The resulting narrow scope of formal, industry level bargaining in Britain initially led to the establishment of joint consultative machinery to supplement the collective bargaining process, a matter that we discuss in the next section. However, in the relatively full employment conditions of the post World War Two years it also produced a demand for a second tier of bargaining at the workplace level. The important point about this workplace level of bargaining, which became so much a focus of public policy discussion in the 1960s, was that it largely concerned itself with subjects which were not covered by industry level bargaining. And prominent among these subjects were health and safety matters. This fact was well evidenced by one of the research papers (33) prepared for the Donovan Commission which reported that 54 per cent and 40 per cent of their shop steward sample claimed to 'discuss and settle as standard practice' safety and health questions respectively; only the broader category of 'general conditions in the workplace' was discussed and settled more often than these two sets of matters. More recently, the 1972 Workplace Industrial Relations Survey reported that 50 and 27 per cent of their shop steward sample respectively settled safety and health questions, with again only general conditions in the workplace (54 per cent) being a more important subject in this regard.(34) Although we have no in-depth knowledge of the exact nature of this shop steward bargaining over health and safety matters it would seem likely to have involved the negotiation of risk differentials, which we discussed in the previous section.

In the years that have followed the report of the Donovan Commission there have been a number of surveys (35) highlighting the movement in the direction of the Commission's recommendations - i.e. towards a system of plant, and to a lesser extent company, level bargaining. On the basis of the argument presented here this movement should result in an increase in the subject matter or scope of formal collective bargaining in Britain, which was an effect that the Commission itself hoped would result from following their basic recommendation. This movement should make possible the increasing negotiation of health and safety provisions of both a preventative and compensatory nature. Although we have no comprehensive evidence on the extent of change in this regard one specific compensatory provision that we can examine here is the coverage of employer sickness benefit schemes. Accordingly, we set out below in Table 3.1 the extent of male manual worker coverage under such schemes at the 2 digit industry level in the manufacturing sector for the year 1970.

The contents of Table 3.1 reveal considerable inter-industry variation in the extent of sickness benefit scheme coverage, ranging from a low of 19.6 per cent in metal manufacture to a high of 92 per cent in coal and petroleum products. Although we make no attempt here to develop and test a comprehensive model designed to try and account for the above variation, we did consider whether there was any systematic relationship between these coverage figures and (i) industrial accident rates and (ii) collective agreement coverage for male manual workers. The resulting Pearson correlation coefficients were -0.051 and +0.036 respectively, which are nowhere near statistical significance, thus indicating that the extent of such coverage could not be seen as compensating for above average accident risk or as deriving from the presence of union power as proxied by collective agreement coverage.

The New Earnings Survey did not provide any detailed information on the <u>nature</u> of such sick pay schemes, but some useful information along these lines was provided by a special study by Incomes Data Services.(36) This study found that compensation for absence due to industrial injury may be incorporated into a sick pay scheme, or may be pro-

Table 3.1: Percentage of Male Manual Employees Covered by Sick Pay Schemes at the SIC Level in Manufacturing, 1970

Industry	Percentage Covered
Food, drink and tobacco	85.6
Coal and petroleum	92.0
Chemicals and allied	85.3
Metal manufacture	19.6
Mechanical engineering	42.3
Instrument engineering	68.9
Electrical engineering	60.7
Shipbuilding and marine engineering	46.9
Vehicles	47.1
Metal goods not elsewhere specified	32.3
Textiles	36.8
Leather, leather goods and fur	27.5
Clothing and footwear	32.4
Bricks, pottery, glass, cement	46.7
Timber, furniture	49.4
Paper, printing and publishing	58.4
Other manufacturing	43.5
All manufacturing	49.0

Source: New Earnings Survey, 1970, Table 112, p.199

-vided by a similar but separate set of arrangements. In cases where they were incorporated in the normal sick pay scheme the terms for absence due to industrial injury were often more favourable than those for sickness. The service qualification may, for example, be waived in the case of industrial injury, or payments for industrial injury not counted against the normal entitlement for sickness absence. Most of the schemes examined made no reference to common law claims, apart from stating that where an accident occurred and a claim is made against a third party no benefit will be paid. In addition to these schemes this study found that some companies had insurance policies to compensate employees for accidents at work. The TUC in their evidence to the Pearson Committee of Inquiry into compensation for personal injury specifically called for improved terms and coverage of employee sickness benefit schemes. This type of demand has also been very prominent in union staff status demands for manual employees. A survey by the Department of Health and Social Security in 1974(37) revealed that there had been a considerable increase in the proportion of employees covered by sick pay schemes since 1970. The Department of Health and Social Security's survey did not provide a manual/non-manual breakdown for individual industries so that an exact comparison cannot be made with the figures in Table 3.1. However, some indication of the extent of increased coverage in these years is provided by figures for the high accident rate metal manufacturing industry; in 1970 only 31.1 per cent of all male workers were covered, but 4 years later the relevant figure was 68.2 per cent. This latter survey also revealed (38) that for 81.9 per cent of the male manual workers covered by such sick pay arrangements the scheme was recognised as part of the contract of employment; the figures in this regard for male non-manual workers was a remarkably similar 81.6 per cent. However, only 68.6 per cent of the male manual workers covered by such arrangements had the scheme recognised as part of the contract of employment and agreed with employee representatives; the relevant figure for male non-manuals was 53.1 per cent. In short, both figures suggest that there is still considerable scope for joint negotiation over the coverage and provisions of sick pay schemes. The General Household Survey for 1976 also revealed substantial differentials in sick pay coverage between occupations and industry. (39) This survey reported that coverage rates among manual workers, which were generally much lower than among non-manuals, had nevertheless improved substantially between 1971 and 1976: from 48 to 60 per cent for skilled workers and from 50 to 60 per cent for semi-skilled workers. There is likely to be even greater scope for such negotiation in relation to other health and safety provisions of both a compensatory and preventative nature. The extent to which this is likely to come about as a result of the provisions of the 1974 Act is a matter considered in later chapters.

The Voluntary Establishment of Joint Health and Safety Committees

As we indicated in Chapter 2, the high accident rate in the coalmining industry has invariably made it a special case, in terms of the early extent of employee and union involvement in health and safety, in a number of countries. For example, a demand for workers inspectors had been put forward at international conferences in 1894, 1896, 1898, 1903 and 1907.(40) In Britain an Act of 1872 gave coalminers the right to appoint a full time employee as a workers inspector with the duty of monitoring safety performance. In 1911 ex-miners became eligible for appointment thus opening up the possibility of full time trade union officers being appointed as inspectors. The Mines and Quarries Act 1954 confirmed and extended these arrangements to the rest of the mining and quarrying industry. These workers inspectors, who must have had five years practical experience as miners, are appointed by the union(s) and are entitled to inspect a mine at least once a month and to investigate all reportable accidents and dangerous occurrences. The inspectors may examine any relevant documents that management is required to keep under the terms of the 1954 Act, and may be accompanied on their inspections by technical experts. (The NUM have appointed a number of professional mining engineers to its staff.) The reports of the inspectors are entered in a book at the mine concerned, with copies being sent to the mines inspectorate and to the union(s). The evidence of the Department of Trade and Industry to the Robens Committee indicated that in 1969 some 6,166 inspections were carried out in 438 coalmines. The position appears to be much less satisfactory in the rest of the industry, with only 354 inspections being carried out at 4,329 quarries in the same year.(41)

However, outside the mining industry the only statutory provisions worthy of brief note are those that provide, under particular regulations, for the appointment by the employer of an employee representative to supervise safety. These particular regulations cover potteries, celluloid manufacture, shipbuilding and building.(42) However, as Williams has noted,(43)

> These are the only examples under the Factories Act of specific provisions for the appointment of safety representatives. There is nothing to compare with the provisions under the mining statutes for the appointment by the workers of their own safety inspectors. Under the Factories Acts the safety representatives are all appointed by the employers... Only the provisions under the Potteries regulations define the powers of the safety representative in detail and require regular inspection with published reports of breaches found and action taken. Following these provisions official policy has moved away from this type of defin-

> ition, which offered some prospect of efficient action, to a much less precise form of supervisory function. The trend has been away from workers participation in accident prevention on a statutory basis. The example established in the mining industry as long ago as 1872 has not been followed.

The above quote indicates that outside the mining industry union involvement in accident prevention, and in workplace health and safety matters more generally, has long been possible only on a voluntary basis. The primary means of this involvement has been through the medium of joint health and safety committees, which various public bodies have attempted to encourage from the early decades of this century. The Factory Inspectorate, for example argued as early as 1913 that,(44)

> ... the experience of several British and American firms show that, in addition to legal safeguards, reduction of accidents can best be secured by obtaining the interest and co-operation of operatives and officials through safety committees. The number and constitution of such committees will depend on the size of the factory and the nature of the industry ... the duties of these committees are to study the causes of accidents, to suggest and devise suitable means for preventing them, to keep careful records, to make frequent inspection of machinery and plant, and to note any defects and dangers. After some experience the principal safety committee usually drafts a code of safety rules applicable to the particular factory ...

In addition to the continual urging of the Factory Inspectorate, the TUC and the central employer body at the particular time have also issued joint statements in favour of the voluntary establishment of such committees.(45) However, despite all this verbal encouragement the extent of progress in the voluntary establishment of joint health and safety committees was for long considered unsatisfactory. This contention is evidenced, for example, by the fact that in 1927 the chief factory inspector introduced a draft order providing for compulsory joint health and safety committees in the high accident rate industries. However, this prospect of compulsion led to such employer opposition that the order was withdrawn on the understanding that employers would accelerate the rate at which committees were being established on a voluntary basis.(46) But some thirty years later a report by the National Joint Advisory Council of the Ministry of Labour still revealed the limited and uneven establishment of such committees: there were "... committees of some kind with functions covering safety ... in about 60 per cent of factories with more than 500 employees, in about 25 per cent of factories with between 250 and 500 workers, in less than 10 per cent of factories employing between 100 and 250 workers, and were almost non-existent in factories with less than 100 employees."(47)

This evidence of the lack of progress in establishing such committees, combined with the increasing accident rate over the period 1963-69, led, as we shall see in the next chapter, to union and parliamentary demands for changes to the Factories Acts which would provide for statutory based joint health and safety committees. These demands did not lead to any immediate legislative changes, although they did result in two surveys by the Factory Inspectorate in 1967 and 1969 which were designed to provide detailed information on the extent of committee establishment.(48) The latter survey, for example, revealed that in total only 27 per cent of the plants investigated had a joint health and safety committee, although a further 20 per cent did have a general consultative committee that discussed health and safety matters. There was also found to be considerable inter-industry variation in the extent to which joint health and safety committees had been established; the range being from a low of 9 per cent of plants in clothing and footwear to 56 per cent of establishments in gas, electricity and water. The findings of these two surveys were considered the most comprehensive and detailed at the time and were in fact presented by the Department of Employment in evidence to the Robens Committee. In addition to these survey returns the Ministry of Labour published in 1968 a study of the detailed workings of 20 such joint health and safety committees.(49) This study indicated that these committees typically meet once a month and tended to discuss accidents and incidents which might have caused accidents, the safety of plant and machinery, gangways and the conditions of floors, methods of material handling, fire precautions and the use of protective clothing and footwear. It was found that the members of these committees had rarely received any specialised training, and that there was very little feedback of the committees' workings and decisions to the work-force at large. These committees were found to be difficult to operate where the general climate of union-management relations was unsatisfactory, and a number of more specific problems, such as poor communication, dissatisfaction with the balance of representation and inadequate provision of necessary information, were also revealed. The committees included in this study were seen to be very much advisory bodies in that they were rarely allowed to take initiatives, such as the investigation of accidents at the time of their occurrence or the conducting of inspections of the workplace on anything like a regular basis. Such functions were very much the responsibility of management appointed safety officers.

The nature of the returns of the Factory Inspectorate surveys provide researchers with little opportunity to try and identify any systematic influences on the decision to establish a joint health and safety committee. This deficiency has, however, been rectified by the recent workplace industrial relations survey carried out by the SSRC Industrial Relations Research Unit at the University of Warwick. The results of this survey, which was conducted in December 1977 and January 1978, provides us with information on the existence of joint health and safety committees established prior to the publication of the Health and Safety at Work Bill. This information, which covers 970 establishments in the manufacturing sector, provides us with the opportunity to try and identify any systematic influences on the likelihood of a plant having a voluntarily established health and safety committee. The basic results of the Survey revealed that in total only 44.4 per cent of the 970 establishments had joint health and safety committees _prior_ to the passage of the 1974 Bill. The results for the individual industries included in the survey are set out in Table 3.2.

The contents of Table 3.2 reveal that only six of the 17 industry groups listed contain more than 50 per cent of establishments having joint health and safety committees prior to the 1974 Bill. Furthermore, if we take only those industries with a relatively sizeable number of sample observations (i.e. >20) then the extent of committee establishment varied between 59.2 per cent of establishments in metal manufacture and 17.9 per cent of establish-

Table 3.2: Voluntarily Established Joint Health and Safety Committees by Industry

Industry	Percentage having Voluntarily established Committees	Total Establishments
Food, drink and tobacco	40.0	110
Coal and petroleum	40.0	5
Chemicals and allied	50.7	69
Metal manufacture	59.2	76
Mechanical engineering	45.8	96
Instrument engineering	52.9	17
Electrical engineering	38.5	96
Shipbuilding and marine engineering	81.8	11
Vehicles	53.7	67
Metal goods not elsewhere specified	39.0	95
Textiles	53.0	66
Leather, leather goods and fur	18.2	11
Clothing and footwear	17.9	67
Bricks, pottery, glass and cement	44.0	25
Timber, furniture	40.6	32
Paper, printing and publishing	47.1	87
Other manufacturing	47.5	40

ments in the clothing and footwear industry. This sort of variation poses the basic question that we are concerned with here, namely why does Plant A have a voluntarily established joint health and safety committee, and not Plant B?

There has been no systematic model developed and tested on a large scale body of survey data to try and explain inter-establishment (inter-industry, etc.) variation in the existence of union-management problem solving arrangements, such as joint health and safety committees. As a result the variables entered in our estimating equation were chosen on the basis of certain rather *ad hoc* hypotheses about the likely factors involved. However, in specifying our set of independent variables we did obtain considerable guidance from (i) a paper by Kochan and Dyer(50) which set out a detailed conceptual framework for analysing the establishment, maintenance and effectiveness of joint organisational change initiatives such as health and safety committees, (ii) a study by Wendling(51) of variation in the presence of health and safety clauses in collective agreements in the United States manufacturing sector, and (iii) a number of case studies of joint consultative arrangements and structures in Britain, in particular that conducted by Derber.(52)

The approaches and findings of the above studies lead us to construct a model that contained three sub-vectors of variables.(53) The first sub-vector sought to account for variation in workplace health and safety hazards. The basic rationale for this sub-vector was the view that joint health and safety committees are adaptive structures, representing an institutional response to a perceived problem. Accordingly, we entered as variables the mean industry reported accident rate (at the MLH or 3 digit industry level) and three technology dummies (continuous process; large batch and mass production; small batch). In the case of the latter variable(s) we hypothesised that large batch and mass production technology has been held to be positively associated with worker 'discontent' as manifested in ways such as high levels of absence, sickness and non-reportable accidents.

The simple existence of a problem, such as that of industrial accidents, does not necessarily guarantee the establishment of joint change initiatives such as health and safety committees. Indeed the history of attempts to encourage the voluntary establishment of such committees reveals, as we have seen, a picture of relatively limited and uneven development, largely as a result of considerable employer opposition to these arrangements on the grounds that they would limit managerial discretion and encourage a proliferation of worker grievances that would interfere with the production process. This sort of evidence suggests that unions will have to have exerted pressure or power in favour of the establishment of such committees. Accordingly, this perspective suggests that one can usefully view the existence of a voluntarily established joint health and safety committee as an institutional outcome of the presence of union power. We therefore included under this second sub-vector the following variables: union density, multi-unionism, the extent of shop steward organisation and variation in industrial conflict, with all these variables being hypothesized to be positively associated with the presence of a joint health and safety committee.

The basic line of argument developed to date is that the voluntary establishment of a joint health and safety committee is the institutional outcome of union power pressuring management to establish such a committee in preference, for example, to simply appointing a safety officer in response to a problem of workplace health and safety. It is almost certainly the case, however, that not all managements will be equally strongly opposed to the establishment of such a committee. Indeed following the argument of Derber we contend that those managements that accord personnel matters a relatively high priority in their decision making calculus will be relatively disposed towards the establishment of joint health and safety committees. As proxies for the personnel orientation of management we therefore entered as variables establishment size, the existence of a general purpose consultative committee, foreign ownership and the presence of a member of senior management with specialist responsibility for industrial relations and personnel matters.

These thirteen variables were then entered stepwise in a discriminant function which sought to distinguish between those establishments with a voluntarily established joint health and safety committee (i.e. one in existence prior to the publication of the Health and Safety at Work Bill) and those without such committees. The results obtained indicated that such committees were positively associated with establishment size, the existence of a general consultative committee, multi-unionism, higher accident rates and a greater number of shop stewards. In terms of our basic sub-vectors of explanatory variables we therefore find some support for the notion that these committees represent an adaptive response to the presence of higher industrial accident rates, but that the bringing about and maintenance of this particular response (as opposed, for example, to solely appointing a management safety officer) does require the presence of some organisational elements of union power, (i.e. multi-unionism and a well developed shop steward organisation) together with a management organisation that accords a relatively high degree of importance to personnel considerations, (i.e. size and presence of a general consultative committee). In both Chapters 4 and 6 we return to this basic framework of analysis to see whether it can help identify the relevant characteristics of firms that have rapidly appointed safety representatives and set up joint health and safety committees following the publication of the Health and Safety at Work Bill.

Conclusions

The general line of argument put forward in this

chapter was that prior to the passage of the 1974 Act the union response to the problem of industrial accidents, and workplace health and safety problems more generally, was disproportionately concentrated on the compensatory, rather than preventative, side of the subject area. A number of reasons were given for this imbalanced response, among which was the constraining force of the legal environment. It is this particular constraint that we examine in the next chapter.

NOTES

1. See, for example, Chris Baker and Peter Caldwell, Unions and Change since 1945, Pan Trade Union Studies, London 1981, Chapter 6.
2. John L Williams, Accidents and Ill Health at Work, Staple Press, London, 1960, p.321.
3. Williams, op.cit. p.322.
4. Geoff Latta and Roy Lewis, "Trade Union Legal Services", British Journal of Industrial Relations, Vol.XII, No. 1, March 1974.
5. See, for example, Kip Viscusi, "Job Hazards and Worker Quit Rates: An Analysis of Adaptive Worker Behaviour", International Economic Review, Vol.20, No.1, February 1979.
6. R B Freeman, "Individual Mobility and Union Voice in the Labor Market", American Economic Review Papers and Proceedings, May 1976.
7. Sidney and Beatrice Webb, Industrial Democracy, Longmans, London, 1902, p.360-61.
8. Annual Report of the Trades Union Congress 1974
9. See, for example, V L Allen, Trade Unions and the Government, Longman, London 1960.
10. Annual Report of the Trades Union Congress, 1972, p.387-8.
11. A Fox, A History of the National Union of Boot and Shoe Operatives 1874-1957, Blackwell, 1958, p.301.
12. Latta and Lewis, op.cit., p.64
13. Latta and Lewis, ibid, p.59
14. R A Hasson, "The Pearson Report", British Journal of Law and Society, Vol.6, No. 1, Summer 1979.
15. Latta and Lewis, ibid, p.64.
16. See, for example, Annual Report of the Trades Union Congress, 1964, p.425.
17. Adam Smith, The Wealth of Nations, Random House Edition, New York, 1937, p.100
18. Robert S Smith, "Compensating Wage Differentials and Public Policy: A Review", Industrial and Labor Relations Review, Vol.32, No. 3, April 1979.

19. Cento G Veljanovski, "The Economics of Job Safety Regulation: Theory and Evidence: Part I - The Market and Common Law", Centre for Socio-Legal Studies, Wolfson College, Oxford, September 1978.

20. Craig A Olsen, "An Analysis of Wage Differentials Received by Workers in Dangerous Jobs", Journal of Human Resources, Vol. XVI, No.2, 1981.

21. John Hicks, The Theory of Wages, Second Edition, 1963, p.110-11.

22. Richard A Zeckhauser, "Procedures for Valuing Lives", Public Policy, 1975, p.455-56.

23. John Mendeloff, Regulating Safety: An Economic and Political Analysis of Occupational Safety and Health Policy, MIT Press, Cambridge, Mass. 1979, p.11.

24. In addition to the reference cited in footnote 20, see G J Duncan and F P Stafford, "Do Union Members Receive Compensatory Wage Differentials?", American Economic Review, Vol.70, No.3, June 1980.

25. See, for example, ASTMS, Guide to Health and Safety at Work, p.2.

26. TUC, Good Industrial Reltions: A Guide for Negotiators, 1971, p.11.

27. David Lewis, "Workers Participation in Safety: An Industrial Relations Approach", Industrial Law Journal, 1974, p.100.

28. Department of Employment Gazette, February 1976.

29. Webb, op. cit., p.357.

30. Arnold R Weber, "Stability and Change in the Structure of Collective Bargaining", in Lloyd Ulman (ed.), Challenges to Collective Bargaining, Prentice Hall, New Jersey, 1967, p.17.

31. Webb, ibid., p.358-60.

32. See, for example, Arthur Marsh, Managers and Shop Stewards, Institute of Personnel Management, London, 1968, p.11-12.

33. W E J McCarthy and S R Parker, "Shop Stewards and Workshop Relations", Research Paper No.10, Royal Commission on Trade Unions and Employers Associations, HMSO, London, 1968, Table 8, p.83.

34. Stanley Parker, Workplace Industrial Relations, 1972, OPCS, 1974, Table 33, p.22.

35. See, for example, D R Deaton and P B Beaumont, "The Determinants of Bargaining Structure: Some Large Scale Survey Evidence for Britain", British Journal of Industrial Relations, July 1980.

36. IDS Study No. 103, Accidents: Pay and Prevention, August 1975.

37. DHSS, Report on a Survey of Occupational Sick Pay Schemes, HMSO, London, 1977, p.XVI-XIX.

38. DHSS, *op. cit.*, p.55-56.

39. OPCS, *General Household Survey 1976*, HMSO, London, p.53-54.

40. Williams, *ibid.*, p.153.

41. Department of Trade and Industry Evidence to the Robens Committee, p.378-9.

42. Williams, *ibid.*, p.165-70.

43. Willaims, *ibid.*, p.171

44. Quoted in Williams, *ibid.*, p.188

45. See, for example, *Annual Report of the Trades Union Congress*, 1963, p.169-70.

46. See Williams, *ibid.*, p.177.

47. Cited in Williams, *ibid.*, p.197.

48. For the details of these surveys see P B Beaumont, *Safety Legislation: The Trade Union Response*, Occasional Papers in Industrial Relations, University of Leeds and Nottingham, No.4, 1979, p.12-14.

49. Industrial Safety Advisory Council, *Works Safety Committees in Practice: Some Case Studies*, HMSO, London, 1968.

50. Thomas A Kochan and Lee Dyer, "A Model of Organisational Change in the Context of Union-Management Relations", *Journal of Applied Behavioural Science*, Vol. 12, No.1, February 1976.

51. Wayne Wendling, "Industrial Safety and Collective Bargaining", *Proceedings of the Industrial Relations Research Association*, Winter 1977, p.427-36.

52. Milton Derber, *Labour-Management Relations at the Plant Level Under Industry Wide Bargaining*, University of Illinois Press, 1955, p.66-80.

53. The full details of the model and results are presented in P B Beaumont and D R Deaton, "Organisational Change in a Union-Management Context: The Voluntary Establishment of Joint Health and Safety Committees in Britain", SSRC Industrial Relations Research Unit, University of Warwick, Mimeographed paper, 1980.

Chapter Four

THE LAW AND UNION INVOLVEMENT IN WORKPLACE HEALTH AND SAFETY

In the previous chapter we referred to the notion that the nature of the legal environment has long constituted an important constraint on the ability of unions to be 'heavily' involved in the preventative side of workplace health and safety. Accordingly in this chapter we consider the nature of these legal constraints, the demands for their removal by the unions and the opportunity for union involvement in the preventative aspects of workplace health and safety provided by the provisions of the 1974 Act. However, before undertaking this task we briefly outline in the following section the case for some form of Government intervention in the market place for workplace health and safety.

The Case for Government Intervention in Workplace Health and Safety

The basic rationale for a Government role in the area of workplace health and safety is provided by the belief that the market mechanism will not produce a socially efficient level of industrial accidents (and disease) and accident (and disease) compensation.(1) This belief is based on the alleged non-fulfilment of the following assumptions: that firms have sufficient knowledge about safety technologies to ensure that they choose the most cost effective methods of accident prevention; and secondly, that the firms bear all the costs of the injuries and illnesses to their employees which result from their production processes. The non-fulfilment of the first assumption means that employers will over-estimate the costs (or under-estimate the benefits) of accident prevention, thus providing a level of safety that is below the social optimal one. This possibility provides a potential role for the Government in disseminating information about prevention measures and in funding or conducting research into new possible prevention measures. In the case of the second assumption, if firms do not fully internalise all the costs of injuries and diseases that result from their production process then the prices of their products will not reflect the full social costs of production. This view of industrial accidents and disease as negative externalities from the production process raises questions about the adequacy of the wage mechanism for compensating workers exposed to above average on the job risks and hazards. If workers, or at least the marginal worker, accurately perceives the risks of industrial accidents and disease, and if the process of bargaining with employers is a relatively cost-less one, then a socially optimal level of safety production can be achieved. However as we saw in the previous chapter, considerable doubt has been expressed about the adequacy of wage hazard premiums in fully compensating workers for the extent of on the job risk.

The belief that the costs of industrial accidents and disease are not fully internalised by firms has provided the basis for Government intervention in the market for workplace health and safety that goes beyond simply the dissemination of knowledge about, and the funding of research into, new accident prevention measures. The overall goal of public policy in the area of workplace health and safety has been argued to be that of seeking to minimise the sum of accident occurrence and accident prevention costs.(2) This perspective, which assumes that the consequences of accidents can be fully and adequately expressed in monetary terms, holds that there is a trade off between the losses arising from industrial accidents and the resources expended to prevent such accidents. According to this line of argument, some positive level of industrial accidents is seen not only to be inevitable, but also to be desirable, at least on economic efficiency grounds, as the costs of such accidents are less than the costs of preventing them. The individual instruments of Government policy in the area of workplace health and safety have been concerned to (i) reduce the risk of accidents, and (ii) to financially compensate the victims of industrial accidents and disease. The concern of this chapter is very much with the former set of measures, although it is important to note the alleged positive, interdependence between the preventative and compensatory sides of Government policy in the area of workplace health and safety. Under systems of common law liability and workmen's compensation, for example, it has long been argued that adequate financial damages and compensation for injured workers should provide employers with a strong financial incentive to reduce the level of accidents. This positive, 'spill-over' effect has, however, been increasingly questioned in the academic literature in recent years.(3) The Report of the Robens Committee also suggested that the legal complexities surrounding actions for damages for personal injury had in practice operated against rather than in favour of accident prevention efforts. (4)

The Regulatory, Preventative Approach in Britain

At the time of the Robens Committee Report there were five major Acts, with supporting orders and regulations, concerned with workplace health and safety. The Factories Act 1961, The Offices, Shops and Railways Premises Act 1963, The Mines and Quarries Act 1954, the Agriculture (Poisonous Substances) Act 1952 and the Agriculture (Safety, Health and Welfare Provisions) Act 1956. Other relevant legislation included a number of Acts which provided for special controls over certain specified industrial activities and substances, such as the Explosives Act, and Acts dealing with emissions and effluents from workplaces. These Acts were supported by over 500 subordinate statutory instruments which have been continually added to; for example, over 100 statutory instruments had been issued under the Factories Act between 1960 and the time of the Robens Committee Report. In Britain industrial legislation setting legal standards for safe working conditions stretches back to the early years of the 19th century, with the pressures and motives for the passage of these Acts being a subject much debated among historians.(5) This body of legislation largely evolved in the following manner,(6)

> This century of experiment in factory legislation affords a typical example of English practical empiricism. We began with no abstract theory of social justice or the rights of man. We seem always to have been incapable even of taking a general view of the subject we were legislating upon. Each successive statute aimed at remedying a single ascertained evil. It was in vain that objectors argued that other evils, no more defensible, existed in other

trades or amongst other classes, or with persons of ages other than those to which the particular Bill applied. Neither logic nor consistency, neither the over-nice consideration of even-handed justice nor the quixotic appeal of a general humanitarianism, was permitted to stand in the way of a practical remedy proved wrong.

The first Factory Act was that of 1802 which sought to limit the working hours of apprentices in cotton mills, and to lay down very general standards of heating, lighting and ventilation. Further Acts in 1819, 1825 and 1831 sought to reinforce or extend these controls. The early development of factory legislation was very much caught up with the shorter hours movement and Acts in this area were passed in 1831 and 1833. A further seven Acts were passed between 1844 and 1856 which imposed a variety of health and safety requirements for different types of mills, and varying limits on the hours of work of children according to age and size of establishment. In 1864 statutory protection was extended beyond the textile industries to six new trades, while in 1867 some heavy industrial processes were brought under the sanitary provisions of earlier legislation, as were all manufacturing establishments employing fifty or more employees. The consolidating Factory and Workshop Act of 1878 ended the division between large and smaller sized establishments by providing a general definition of factory premises which, to a very large extent, remains operative in present day legislation. Subsequent factory legislation consisted of further detailed regulation with periodic attempts at consolidation. In this piecemeal, _ad hoc_ development of workplace health and safety regulations the major step forward is often seen to be the 1833 Factory Act which provided for the appointment of four inspectors to enforce the Act's provisions. Indeed Wedderburn has gone so far as to argue that "those four inspectors were perhaps the most important innovation in British labour legislation."(7) This establishment of an inspectorate was seen as a necessary if not sufficient condition for effective factory legislation, and administrative inspection and criminal prosecution has remained the main method of enforcing present day health and safety laws and regulations. The value of the inspectorate function in the area of workplace health and safety has, however, come in for increasing criticism on a variety of grounds. For example, an article by Greenberg(8) examined various measures of the 'activity rate' of the Factory Inspectorate for the period 1950-69. The basic (mean) statistics from this study are set out below in Table 4.1.

This study failed to find any meaningful correlation between the various activity measures of the work of the Factory Inspectorate and movements in the industrial accident statistics - i.e. there was no indication that the Factory Inspectorate had favourably influenced the level of industrial accidents, or even that its activity rates adjusted themselves to follow up any adverse movement in industrial accidents. The basic conclusion reached was that the inspectorate tended to operate in a vacuum, taking little account of changes in the parameters of workplace health and safety. This sort of conclusion was reflected in the Robens Committee recommendation that "the resources of the inspectorate must be used selectively. They should be concentrated on priorities and problems that have been identified through the systematic assessment of all the available data."(9) There have been other studies which have also questioned the value of the inspection function through demonstrating considerable differences between the inspectors' statutory duties and their working practices.(10) In this regard the particular aspect of the inspectorate that has come in for most adverse comment is its reluctance to prosecute employers found to have infringed the relevant regulations.(11) The day to day emphasis of factory inspectors on an educative, rather than prosecution, role is certainly not unique to this Inspectorate;(12) indeed Hartley(13) has argued that it is an almost inevitable fact of life where any understaffed inspectorate is seeking to ensure compliance with a rather unclear set of legal regulations.

More generally, the Robens Committee argued that the ad hoc development of workplace health and safety regulations in Britain had produced a situation of too much law and an excessive fragmentation of administrative jurisdictions. In the committee's view, the result had been counter-productive in that "people are heavily conditioned to think of safety and health and work as in the first and most important instance a matter of detailed rules imposed by external agencies."(14) Their contention was that too much reliance on legal regulation and the external, inspectorate function had discouraged voluntary initiatives concerned to improve health and safety conditions at the individ-

Table 4.1: The Record of the Factory Inspectorate, 1950-69

Variable	Mean Statistic
Visits to Factories	206,060
Visits to other places	35,143
Other visits	49,821
All visits	291,025
Inspectorate staff	456
Physical examinations	811,100
Number of prosecutions for health	208
Number of prosecutions for safety	914
Number of all prosecutions	1,535
Inspectorate expenditures	756,641
Average fine, health suits	5.29
Average fine, safety suits	16.12
Average fine, all suits	11.34
Fatal accidents	682
All accidents	213,883
Dangerous occurrences	1,322
Occupational diseases	475
All mishaps	215,680

Source: L Greenberg, "Does Government Enforcement Help Industrial Safety?", Engineering, September, 1972, p.855.

ual workplace. This outcome was largely seen to have resulted from a psychological conditioning effect: "the existence of such a mass of law has an unfortunate and all-pervading psychological effect." (15) This contention is taken a stage further in the next section where we seek to demonstrate that the very nature of the external, legal regulations prior to the 1974 Act seriously constrained any substantial union involvement in the area of workplace health and safety.

The Legal Constraints on Union Activity and Demands for their Reform

The employers right of control of the workforce was and still is central to the whole enforcement system of workplace health and safety and imposes on the employer a duty to warn, instruct and discipline his employees in the interests of health and safety. This duty to warn, however, went no further than was necessary for the requirements of immediate safety, and could not be expanded into a duty to inform and consult the workforce upon the general hazards of the processes.(16) This placing of the basic responsibility for workplace health and safety in the hands of management has long made for a highly paternalistic common law approach to the subject. Furthermore, "the statutory safety codes, which developed from these common law principles, were from the outset equally paternalistic in concept and enforcement: while no doubt intended for the benefit and protection of the worker, they operated over his head rather than through his cooperation." (17) In all of the industries which were subject to these codes, the workers health and safety obligations were derived solely from the concept of avoidance of injury to himself and others. As the Department of Employment put it, in their evidence to the Robens Committee, "their (employed persons) responsibility is to refrain from interfering with the various arrangements made for their safety; to use any equipment provided for their protection; to behave with due regard to their own safety and that of their fellow employees; and to report any dangerous hazards of which they become aware."(18)

The converse of these limited responsibilities, extending little beyond that of workplace discipline, was limited rights in the way of employee and union involvement in the area of workplace health and safety. Under the Factories Act 1961 employees or their trade union representatives had no rights to inspect any of the statutory safety and health records kept at the workplace, no legal right to liaise with a factory inspector on a visit or even to see an inspector's report which may affect them as individuals. More fundamentally, the employees had no formal rights to any information about health and safety hazards under the statutory codes, beyond the relatively limited rights bestowed by the common law. Similarly, no trade union representative had any legal right to information either from the employer or the inspectorate about the hazards to which any of his members might be exposed at work. The sole rights conferred upon trade unions by the Factories Act 1961 were to attend inquests on members (or workers of a class organised by the union), to attend formal inquiries into accidents or hazardous occurrences and, if they represented a majority of the employees concerned, they could lodge an objection to regulations in draft, and thus secure a public inquiry at which the objecting union secured *locus standi*. The much smaller number of workers covered by the Mines and Quarries Act 1954 had considerably more rights than those listed above.

In this case all records under the Act were open to the workers or their representatives, all accidents that occurred had to be notified to both the inspector and the union, all reports of workers inspectors had to be posted at the pithead and, if disputes on good mining practice between an inspector and management were taken to arbitration, the trade union had *locus standi* in the resulting proceedings.

The system of union appointed safety inspectors in the coal mining industry, which was extended to the rest of the mining and quarrying industry by the 1954 Act was, as we saw in the previous chapter, virtually unique in Britain. However, during the course of the 1960s (19) unions in the manufacturing sector began to urge reform of the Factories Act, with a view to establishing similar employee and union rights to those existing under the Mines and Quarries Act 1954. Indeed during this period of time such demands became one of the most debated issues at the annual meetings of the TUC. This came about largely as a result of (i) the increasing evidence of the limited and uneven development of joint health and safety committees on a voluntary basis, and (ii) the rising industrial accident rate in these years; matters that we discussed in the two previous chapters. The nature of the motions that were put forward at congress during this period of time are well illustrated by that presented by a spokesman for the Amalgamated Union of Foundry Workers in 1964.(20)

> Congress declares that in order to provide effective safety organisation in industry, the Factories Act should be amended to provide for:
>
> (a) the election of safety delegates, with powers of inspection, by the workers concerned in factories: such powers of inspection to include the right to inspect the scene of an accident and the equipment involved, a right at present available only to miners under the Mines and Quarries Act 1954;
>
> (b) the setting up of safety committees in factories;
>
> (c) the right of workers' safety delegates to accompany the factory inspector on his visits to factories;
>
> (d) the advice of the factory inspector to the firm to be available to the safety commit-

tee or safety delegates.

The general council of the TUC initially opposed the above motion, largely on the grounds that statutory joint health and safety committees might possibly reduce employer liability for accidents.(21) However, in the following year a representative of the general council conceded that the voluntary approach to establishing joint health and safety committees had failed and that it was therefore necessary to legislate for such arrangements.(22) The TUC, which was considerably influenced by the statutory based system of worker safety delegates and joint safety committees in Sweden, began to push for the introduction of statutory based arrangements from about the mid-1960s, but the Government made little initial movement in this direction. Indeed in 1966 it refused to support a Private Members Bill for statutory joint safety committees, although it did announce that if satisfactory progress in setting up joint safety committees on a voluntary basis was not apparent in the next few years it would feel obliged to seek the power, through reforms of the Factories Act 1961, to require joint consultative machinery to be set up in appropriate cases.(23) It was this statement that led to the Factory Inspectorate surveys of joint health and safety committees in 1967 and 1969 that we discussed in the previous chapter.

The 'limited' progress revealed by these two surveys led to the introduction of the Employed Persons (Health and Safety) Bill in February 1970. This Bill proposed to replace the 'factory doctors' appointed under the Factories Act with a new Employment Medical Advisory Service (this subsequently occurred in 1973), and the appointment, by a trade union 'recognised for the purpose of negotiating terms and conditions of employment', of safety representatives at all factories with more than 10 employees. These safety representatives were to have the right to make inspections every 3 months, or more frequently in the case of accidents, and were to have access to all relevant documents which the employer was required to keep under the terms of the Factories Act. These safety representatives were to be persons with five years employment in the industry, over 23 years of age and if possible with two years employment at the factory concerned. In all factories with more than 100 employees management was to be required to set up a joint safety committee if requested to do so by the safety representatives.

However, Parliament was dissolved in June 1970 with the Bill not reaching the statute book and the Conservative Government which subsequently took office refused to carry on with it. Instead they stated their intention to await the Report of the Robens Committee that had been set up by the previous Labour administration. The Report of the Robens Committee conceded that a statutory provision requiring the appointment of safety representatives and safety committees might be advantageous, but refused to recommend such a provision on the grounds that it might prove 'too rigid' or 'too narrow' in concept.(24) Instead the Report proposed a statutory requirement dealing in general terms with arrangements for participation by employees. Following this Report the Conservatives presented a Bill on Health and Safety at work to Parliament in January 1974. This Bill was largely based on the report of the Robens Committee, although it conceded even less in the way of union involvement. However, this Bill was overtaken by the General Election of the following month, and when the Labour Party was returned to power it passed the Health and Safety at Work Act 1974.

The Health and Safety at Work Act 1974: An Enabling Measure

The Health and Safety at Work Act 1974 provides an update of previous health and safety at work statutes and regulations, and also extends statutory safety protection to some five million workers previously outside the orbit of any legislation. The Act is essentially an enabling measure since the new Health and Safety Commission, as the main administrative agency, will make recommendations to the Secretary of State on the issuance of new regulations which will, over time, extend, revise or replace the existing legislation. These regulations will cover a wide range of health and safety matters at the workplace. In addition to these regulations two other forms of obligation are imposed on employers; codes of practice to be issued by the Commission, and general statutory duties on employers concerning their health and safety obligations to their employees and the general public, breach of which is an offence. There is also a statement of the general duties of employees in relation to health and safety matters at work. The basic aim of the Act is to create an essentially self-regulating system for the adoption and observance of safety and health standards and regulations; this approach follows from the Robens Committee's contention that previously too much responsibility for the adoption and observance of workplace health and safety standards had been placed on the law and the inspectorates, and too little on employees and management.

The Health and Safety at Work Act 1974 is similar to earlier safety legislation in placing heavy responsibilities upon management, retaining inspectors to monitor compliance with its provisions and in providing for enforcement in the criminal courts. (25) However, the Act is a significant departure from these earlier Acts in the extent of its provision for employee involvement. The general duty of the employer specifically includes the provision of information to ensure the health and safety of his employees, and the inspectorate has the duty of providing information to employees about the premises where they work and about any action to be taken which concerns them. These duties were immediately brought into force by the Act and have applied to all employed persons since April 1, 1975. However, the most 'radical' departure of the new Act was the entitlement to appoint employee safety representatives who would be entitled to consult with the employer "... with a view to the making and maintenance of arrangements which will enable him and his employees to cooperate effectively in promoting and developing measures to ensure the health and safety at work of his employees, and in checking the effectiveness of such measures." The Act does not precisely specify the duties of the workers safety representatives, but simply states that they shall represent the employees in consultations with management. The Employment Protection Act 1975 amended the Health and Safety at Work Act to give recognised trade unions the sole right to appoint safety representatives. The Employment Protection Act's substitution of the original provision (of the Health and Safety at Work Act) for employee appointed safety representatives with the stipulation that representatives be solely union appointed was the most controversial aspect of the safety representative regulations. This change predictably led to considerable criticism by the Conservative opposition in Parliament and by employer organisations such as the CBI. The essence of this criticism was that industrial relations considerations (namely the Employment Protection Act's explicit intention to foster collective bargaining) had conflicted with, but triumphed over health and safety considerations.

The alleged adverse consequences of this decision have been outlined in the following terms,(26)

> ... apathy towards safety is likely to be greatest in just those establishments which are not unionised, but the Employment Protection Act appears to do little to foster the growth of trade unions in workplaces where there is not already a substantial interest in collective bargaining. It may be permissible to tolerate apathy in industrial relations, but it is arguable that lack of union representation ought not to cut off employees from consultation in respect of safety, particularly if, as may possibly be the case, unsafe working conditions are prevalent in establishments which are not unionised. It would be most unfortunate if the philosophy of satisfactory industrial relations were allowed to conflict with the philosophy of safety at the workplace.

The essence of the above contention is that workplace health and safety problems may be disproportionately concentrated in the non-union sector. In fact, our earlier analysis of the data in Table 1.2 demonstrated that it is union members that actually face a disproportionate amount of such problems, at least in the sense of being very largely employed in the high accident rate industries. The demonstration of this fact means that the decision to restrict safety representatives to being union appointed can be seen to have a health and safety, as well as an industrial relations, basis. However, this finding does not mean that the nature of employee involvement in workplace health and safety matters in non-union establishments is an unimportant issue; indeed we discuss it as an area for future research in our concluding chapter.

In November 1975 the Health and Safety Commission published a consultative document on safety representatives and safety committees. It was expected that the relevant regulations would be brought into operation by May 1, 1976, but both the CBI and TUC made such extensive criticisms of this consultative document (27) that the revised proposals were not published until November 1976.(28) The revised proposals were that regulations on safety representatives and safety committees should be made under section 2(4) of the 1974 Act, prescribing the cases in which recognised trade unions might appoint safety representatives, specifying the functions of such safety representatives and setting out the obligations of employers towards them. A code of practice amplified the regulations to give guidance on how safety representatives should fulfil their statutory functions and guidance to employers regarding the information which they should make available to safety representatives to enable them to fulfil their functions. The guidance notes also offered advice to all who would be concerned with the appointment and functioning of safety representatives, and similarly with regard to safety committees.

Under the regulations a recognised trade union would be able to appoint safety representatives from among the employees of the employer by whom it was recognised. It was envisaged that the number of representatives to be appointed would be determined by agreement between employers and unions, but it was suggested that the relevant criteria should include the total number of employees, the variety of workplace locations, the operation of shift systems, the type of work activity and the degree and character of inherent dangers in the work process. A safety representative should, at the time of his appointment, desirably have been employed by his employer throughout the preceding two years, or have at least two years experience in similar employment.

The functions of safety representatives would be primarily to consult with employers in accordance with Section 2(6) of the Act, but a representative should also investigate potential hazards, dangerous occurrences and complaints by any employees he represented, and make representations about these matters to the employer. He would make representations to the employer on general matters affecting the health, safety or welfare at work of the employees, carry out inspections, represent employees in consultation with the inspectorate, and receive information from the inspectors. It was intended that no function given to a safety representative by the regulations should impose any duty on him, but this protection should be without prejudice to the general duties imposed upon employees and others by the Act. The proposed regulations would provide that the safety representative should be permitted paid time off in order to receive training and to perform his functions; the Health and Safety Commission subsequently issued a separate code on time off for the training of safety representatives.(29)

Safety representatives would be entitled to

inspect the workplace if they had given their employer reasonable written notice of their intention to do so, and had not inspected it in the previous three months. More frequent inspections could be arranged by agreement with the employer or in circumstances where there had been a substantial change in the conditions of work or where an official publication provided new information relevant to the hazards of the workplace since the previous inspection. The employer would be entitled to be present at the inspection and would have to provide such facilities and assistance as the representatives reasonably required. Safety representatives would be entitled to carry out an inspection where there had been a notifiable accident or dangerous occurrence or a notifiable disease, provided they had notified the employer of their intention. Safety representatives would be entitled to inspect copies of documents which the employer is required to keep for the purposes of health and safety legislation. The employer would also be required, with certain safeguards, to make available to safety representatives, information necessary to enable them to fulfil their function. The employer would be required to establish a safety committee within three months where at least two safety representatives requested in writing such a committee. The employer would be required to consult with the representatives as to the composition of the committee. The guidance notes provided advice concerning the objectives, functions, membership and conduct of safety committees. In considering the potential authority and power of safety representatives in Britain a number of useful comparisons have been made with the position of their counterparts in other European countries. These studies have revealed that, in comparison with, for instance, the position in Norway, Sweden and West Germany, the safety representative in Britain appears to have much less extensive rights to information, much less specific provisions with regard to paid time off for training and certainly nothing like the right to stop the work which safety delegates in Norway and Sweden have where they believe there exists an immediate danger to life or health.(30)

The Secretary of State for Employment stated in November 1976 that "it has been decided ... not to bring these regulations into force for the time being ...".(31) This decision was justified on the grounds of the 'high costs' to employers of providing safety representatives with paid time off for training and the performance of their functions, particularly in the public sector where attempts were being made to reduce expenditure and where local authorities had suggested that the cost of paying representatives and administrative overheads could amount to £80 million per annum(32) The Health and Safety Commission were predictably concerned that this delay in implementation would undermine their own credibility as well as that of the regulations themselves. Following representations and discussions the Secretary of State announced that the Government approved the regulations and that they would come into force in October 1978.

The Early Impact of the Safety Representative Regulations

One of the inevitable results of this relatively long drawn out process of implementation (i.e. an actual "lead in" period of 18 months, rather than the 4 months originally proposed) was suggested by the following comment of the Chief Inspector of Factories in his Annual Report for 1977, (33)

> The Safety Representatives and Safety Committees Regulations are in force from 1st October 1978. Therefore Area Directors were not able to comment authoritatively about the way the whole of industry had dealt with the question of safety representatives since there were clearly two schools of thought: those employers who faced squarely up to the issue and those who avoided it until they were legally obliged to accept the reality of safety representation among the workforce.

This relatively early suggestion of variation in employer response to the regulations is an important part of the background to the Health and Safety Executive Survey conducted in October 1979 that was designed to assess the extent to which safety representatives had been appointed under the Safety Representative/Safety Committees Regulations during their first year of operation. It is the basic results of this survey which are analysed in this section.

This survey, which covered some 6,630 workplaces, was conducted by means of a questionnaire administered at those workplaces routinely inspected by the Factories, Mines and Quarries, and Agricultural Inspectorates during the period 1-26 October 1979. In response to the question "Have the safety representative and safety committees regulations been invoked at the establishment by one or more recognised trade unions" the survey obtained the basic distribution of results set out in Table 4.2. As the vast majority of workplaces in the survey had less than 50 employees there were found to be relatively few workplaces where safety representatives had been appointed, i.e. only 17 per cent in total. The industry orders, construction and agriculture accounted for just under 50 per cent of the workplaces surveyed, but only 5 per cent of these workplaces had safety representatives, so that without these two orders the percentage of total establishments with safety representatives rises from 17 to 29 per cent. The influence of small plant size is further revealed by the fact that if we consider numbers of employees covered rather than workplaces covered, then over three-quarters of all employees were represented by a safety representative. Nevertheless, if we ignore for the moment those industries with relatively few sample observations then only six of the 27 industry orders had more than 50 per cent of establishments where the safety representative regulations had been invoked during the period of time under examination. Furthermore, there was very substantial inter-industry variation in this regard, from a low of 0.9 per cent in agriculture and forestry to a high of 66.7 per cent in gas, electricity and water.

Ideally, we would like to try to explain why the safety representative regulations were invoked at Establishment A and not at Establishment B over this period of time using a 'micro data' set, i.e. one that is based directly on individual establishment level characteristics. However, this was not possible as the questionnaire did not seek to obtain any establishment level data beyond that of workplace size. Accordingly, the approach adopted here is to try to explain inter-industry variation at the 2 digit industry level (i.e. column (1) in Table 4.2 is our dependent variable) using a set of industry mean statistics drawn from various external data sources.

The model outlined and tested here had its basis in two pieces of work. First, the analysis presented in chapter 3 that was concerned with the relevant characteristics of establishments that had voluntarily set up joint health and safety committees well before the passage of any legislation on the subject. And secondly, the Daniel and Stilgoe management survey of the impact of industrial relations legislation (34) passed by the previous Labour Government. The approach adopted here can be seen, at least to some extent, as an attempt to (i) organise (under appropriate sub-headings) and (ii) extend the range of variables put forward by Daniel and Stilgoe, who did little more than consider, in a rather *ad hoc* manner, the influence of plant size and the extent of union organisation in accounting for variation in the impact of the legislation at the establishment level. In an attempt to provide a more systematic and comprehensive framework of explanation we put forward a model which consists of the following three sub-vectors of variables: (i) variation in the incidence and extent of concern over the problem involved (in this particular case, health and safety matters); (ii) variation in the extent of union power; and (iii) variation in the personnel orientation of management.(35) These are of course the same basic sub-vectors employed in the analysis of chapter 3.

Under the first sub-heading we seek to account for variation in the extent of concern over health and safety matters. In this regard the first variable that we employed was the industry accident rate which we saw as an important source of concern with health and safety matters. The industry accident rate is important not only in its own right but also because of its positive and significant correlation with other objective characteristics and behavioural manifestations of a poor quality working environment. In addition to the accident rate variable we sought a variable that would constitute an institutional manifestation of the extent of concern with health and safety matters. The variable utilised for this purpose was the extent of the voluntary establishment of joint health and safety committees. These joint health and safety committees, which were established prior to the passage of the Health and Safety at Work Act 1974, are viewed as adaptive structures which represent an institutional response to a perceived problem of workplace health and safety.

The simple existence of a problem of workplace health and safety will not necessarily guarantee a relatively rapid introduction of the safety representative regulations. The fact of the matter is that there are numerous detailed issues and potential problems that have to be resolved in the course of union-management discussions and negotiations before these regulations are likely to come into operation. The perspective adopted here is that

Table 4.2: The Relationship Between the Invoking of the Safety Representative Regulations and the Industry of Employment (%)

Industry	Regulations Invoked (1) YES	(2) NO	Total Number of Establishments
Agriculture and forestry	0.9	99.1	1129
Mining and quarrying	49.7	50.3	314
Food, drink and tobacco	36.9	63.1	141
Coal and petroleum	64.3	35.7	14
Chemical and allied	50.4	49.6	113
Metal manufacture	51.2	48.8	123
Mechanical engineering	22.3	77.7	273
Instrument engineering	6.5	93.5	31
Electrical engineering	32.7	67.3	107
Shipbuilding and marine engineering	20.7	79.3	29
Vehicles	44.6	55.4	83
Metal goods not elsewhere specified	17.7	82.3	356
Textiles	38.2	61.8	110
Leather, leather goods and fur	0.0	100.0	10
Clothing and footwear	11.9	88.1	118
Bricks, pottery, cement and glass	23.1	76.9	108
Timber, furniture, etc.	11.2	88.8	206

Paper, printing and publishing	37.6	62.4	173
Other manufacturing industries	23.0	77.0	152
Construction	3.9	96.1	1912
Gas, electricity and water	66.7	33.3	54
Transport and communication	38.1	61.9	42
Distributive trades	14.2	85.8	148
Insurance, banking and business services	0.0	100.0	3
Professional and scientific services	53.8	46.2	240
Miscellaneous services	7.6	92.4	526
Public administration	60.9	39.1	115
Total	17.2	82.8	6630

union-management agreement over such matters is likely to be substantially facilitated (i.e. such negotiations are likely to be significantly shortened), with the result that the regulations had been invoked during the period of time under examination, in the presence of various sources and manifestations of union power.

The resolution of partially conflicting union-management interests through the exercise of structurally based sources of power is central to the industrial relations paradigm, although the task of devising a fully comprehensive set of empirical proxies for the concept of union power is admittedly a far from straightforward one. The most widely used individual organisational proxy for union power has undoubtedly been the extent of workforce organisation. This particular variable is held to measure the potential elasticity of substitution between organised and unorganised labour. Accordingly, we used collective agreement coverage as a measure of the extent of workforce organisation, hypothesising that the higher collective agreement coverage for male manual workers the more likely that the safety representative regulations would have been invoked in their first year of operation.

Under this union power sub-vector we also need to consider the specific development of shopfloor organisation for collective bargaining purposes. The potential importance of this consideration follows from the general argument that the shopfloor has enjoyed substantial power gains in relation to both management and the official union structure in the relatively full employment environment of the post-war years. In a series of case studies one could seek to measure or proxy the extent of development of shopfloor organisation for collective bargaining purposes by reference to variables such as average shop steward constituency size, the extent of development of hierarchy within this body (e.g. appointment of senior shop stewards), the regularity of steward meetings, etc. However, as these particular variables are rarely available on a systematic industry by industry basis we had to utilise the nature of bargaining structure as a proxy for the extent of development of shopfloor organisation for collective bargaining purposes. The basic contention here was that the greater the extent of single employer bargaining (conducted at the plant and to a lesser extent the company level), as opposed to multi-employer, industry level bargaining, the greater was the extent and nature of union development and power at the shopfloor. And hence the more likely it was that the safety representative regulations would have been invoked during this period.

If one moves away from essentially organisational measures or proxies of union power to take a more behavioural orientated perspective then the particular proxy that has traditionally been utilised for this purpose is the incidence or frequency of strikes. However, this particular variable has a number of potential limitations as a proxy for union power. In this regard one could, for example, argue that it may be the threat of a strike, rather than the actual carrying out of this threat, that is the really important dimension of union power. If this is in fact the case then it is arguable that the unions in a really powerful bargaining position rarely have to go on strike; the simple threat to do so is sufficient to achieve their negotiating aims. A second possible limitation of strike frequency as a proxy for union power is that it does not take account of the fact that union power can be exerted or manifested through other means than strikes, such as go-slows, work to rules, overtime bans, etc. Despite these *a priori* reservations we followed traditional practice and entered strike frequency as a behavioural proxy for union power, predicting a positive sign on the variable. However, we did broaden the notion of the strike as a major instrument of union power by entering as a variable the working days lost dimension of strike activity.

The implicit assumption underlying the sub-vector of union power variables outlined above is that the extent of management opposition to the relatively rapid introduction of the safety representative regulations was essentially a *constant* across establishments and industries. In practice, however, it was almost certainly the case that not all managements were equally opposed to the introduction of these regulations, a fact that has obvious implications for the speed with which union-management agreement over their details was reached. Accordingly, the basic argument underlying the third and final sub-vector of variables outlined here is that those managements that accord personnel matters a relatively high priority in their decision making calculus will have been relatively disposed towards the introduction of the safety representative regulations.

Under this third and final sub-vector we, firstly, entered establishment size as a variable. The use of the size variable as a proxy for the personnel orientation of management follows from the results of a number of studies which have indicated that larger plants devote proportionately more resources to their personnel management function than their smaller sized counterparts, possibly because of the alleged non-pecuniary disadvantages of employment in the more structured, regimented environment of larger plants. As a second potential source of influence we need to take account of the traditional organisational development and change literature which has so strongly stressed the importance of the senior management role in the successful initiation of a change effort. In keeping with our general line of argument concerning the importance of the extent of the personnel orientation of management in this process we hypothesised that where there was a person with a specialist responsibility for industrial relations and personnel matters present on the board or most senior level of decision making within the organisation then there was a significant likelihood of the safety representative regulations having been invoked during the period of time under scrutiny. The basic argument here is that where the personnel management function has attained such relatively high status within the management decision making hierarchy this has the effect of making the individual establishment's management especially conscious of, and responsive to, the provisions of industrial relations legislation.

Finally, we entered a public sector dummy variable, hypothesising that there was a significantly greater likelihood of the regulations having been invoked in industries predominantly or solely in the public sector. This expectation follows from the Government's acceptance of an obligation to act as a 'good employer' of labour in the public sector. This good employer obligation has frequently placed the public sector to the forefront of the process of introducing new institutional arrangements and structures in the industrial relations field in Britain. Accordingly, we hypothesised that this good employer influence in the public sector would have been at work in encouraging a relatively rapid introduction of the safety representative regulations.

The correlation results obtained from this exercise are set out in Table 4.3. Our set of independent variables could be most fully tested (for reasons of data availability) on the manufacturing sector only observations (N = 17) and here the results obtained suggested that the safety represent-

Table 4.3, Correlations between Independent Variables and the Invocation of the Safety Representative Regulations

Independent Variables	Manufacturing Sector only (N = 17)	All Industries /Services (N = 27)
Reported Industry Accident Rate	0.6431*	-
Voluntarily Established Joint Health and Safety Committees	0.3204	-
Collective Agreement Coverage	0.7133*	0.7839*
Single Employer Bargaining	0.5476*	-0.1866
Strike Frequency	0.3029	0.2650
Strike Impact	0.0993	0.0963
Establishment Size	0.7112*	0.5018*
Personnel/Industrial Relations Director	0.5941*	-
Public Sector	-	0.6286*

* Statistically significant

ative regulations were most likely to have been invoked during their first year of operation in industries characterised by high accident rates, high collective agreement coverage, single employer bargaining, large sized establishments and establishments where there was a member of senior management specifically responsible for industrial relations and personnel matters. In the case of the all industries/services set of observations (N = 27) the establishment size and collective agreement coverage variables were again statistically significant, but the individual variable of most particular interest here was the public sector employment dummy. That was in fact positively signed and statistically significant indicating the greater likelihood of the safety representative regulations having been invoked in public sector industries. The strength of the collective agreement coverage variable in particular was confirmed by multiple regression analysis, but too much should not be made of the results of such analysis because of the relatively limited number of industry observations, together with the evidence of substantial multi-collinearity; hence our preference for using correlation analysis.

If one can generalise from the results obtained here it would seem that in the early stages of the introduction of any industrial relations law, which gives new rights to unionised employees, the relevant provisions are likely to be operationalised very largely in establishments with the sort of characteristics indicated by our three basic subvectors of variables. However, through the course of time this process of operationalisation should become increasingly random in nature. This is certainly our expectation with regard to the safety representative regulations, although it would obviously be useful to conduct follow-up surveys to that above in order to identify just when this effect began to come about.

The Early Issues of Discussion and Debate in Relation to the Regulations

In spite of the above-mentioned delay in implementation there was evidence of considerable activity in relation to the regulations, with safety representatives being appointed, joint committees established, or reformed and health and safety training courses going ahead. Indeed in Chapter 6 we utilise further evidence from the SSRC workplace industrial relations survey which was drawn on in the analysis of Chapter 3, as well as from the Health and Safety Executive Survey utilised above to document the extent of establishment of new joint health and safety committees both prior to and immediately after the regulations becoming law in October 1978. The 'reform' of already existing committees under the influence of these 'coming regulations' will also be discussed. It is with the issue of health and safety training courses, however, that we begin the discussion of this final section of the chapter.

The design and implementation of training courses for safety representatives was a major area of union activity prior to the regulations becoming law in October 1978. In May 1977 a TUC conference on health and safety stressed that the emphasis of such training would be on TUC-approved courses only, (36) with the key functions of such training being to help identify health and safety issues in the workplace, find appropriate means and standards for dealing with health and safety problems and help establish an 'infallible union workplace organisation' to ensure that employers actually implemented safety measures. Some individual unions have set up their own courses for safety representatives, but the main training initiative has rested with the TUC. In this regard the TUC has run briefing workshops for full-time officers and produced a course outline to allow these officers in turn to provide a one or two day course on health and safety organisation for their safety representatives. There have also been 5 or 10 day courses, provided on a day release basis at colleges of education, polytechnics and WEA centres. In the first half of 1978 it was estimated that some 1,300 full-time officers had attended 2 day workshops on health and safety, and a further 10,400 workplace representatives had undertaken the 5 to 10 day package training course. The TUC's target was that some 160,000 representatives would have undergone training in health and safety by 1980. Although the TUC failed to attain this extremely ambitious target figure the scale of the safety representative training was undoubtedly impressive. This is evidenced by the figures for the number of health and safety courses and the number of representatives attending these courses which are set out in Table 4.4.

An Incomes Data Services study(37) in early 1980 estimated that approximately half the total number of safety representatives in the country had received training by the end of 1979 and that the level of provision for such training was likely to begin levelling off; the 1979-80 figures in Table 4.4 certainly confirm that expectation. The extent and nature of this training programme was certainly sufficient to cause the Chief Inspector of Factories to go on record as saying that safety representatives were becoming so well trained that they know more about workplace health and safety matters than many of their managers.(38) The issues, implications and impact of this large-scale training programme, which was mounted in a relatively short space of time, and covered relatively more technical subjects than has traditionally been the case in trade union courses, is certainly a sizeable and

Table 4.4: TUC Courses and Students in Health and Safety, United Kingdom 1974-1980

Year	H & S General	H & S Sector	Total Safety	Basic General	Basic Sector	Follow On	Total Courses	H & S as % of total Courses
1974-75	N/A	N/A	35	N/A	N/A	N/A	787	4
1975-76	300	61	361	363	320	104	1148	31
1976-77	394	179	573	472	280	230	1555	37
1977-78	503	238	741	613	410	244	2008	37
1978-79	1021	723	1744	743	362	251	3100	56
1979-80	900	544	1444	804	404	383	3055	48

Year	Health & Safety	Basic Shop Stewards	Follow On	Total Courses	H & S as % of Total
1974-75	N/A	10,904	N/A	10,904	N/A
1975-76	5,360	9,463	1,441	16,264	33
1976-77	7,803	10,917	2,652	21,372	37
1977-78	10,398	14,047	3,034	27,479	38
1978-79	27,361	13,421	3,074	43,856	62
1979-80	18,808	15,701	4,542	39,051	48

Source: Compiled from the Annual Reports of the Trades Union Congress

important subject area for future research, possibly building on some of the insights obtained from previous studies of shop steward training.(39) Finally, on the matter of time off for training it is worth noting that the above study by Incomes Data Services found that, among fifteen companies, there were no separate agreements on time off for safety representatives training, even though the legal rights to paid time off for such representatives are quite different from those for other workplace representatives. Moreover, there has been at least one well published case of an employer refusing permission for one of his safety representatives to attend a union course, on the grounds that an in-company course run by management was provided and hence the request for attendance at the union course was unreasonable.(40) A complaint was made to an industrial tribunal, and on appeal, it was held that in such circumstances it may not be unreasonable for an employer to refuse paid time off to a safety representative.

As background to our survey of safety representatives, the results of which are presented in the next chapter, it is essential to consider some of the specific issues that have been prominent in union discussions about the safety representative and safety committee regulations. The first such issue has been the question of whether or not the safety representatives should be shop stewards. This question provoked different responses from different unions. Some unions like the AUEW (engineering section) and NALGO specified that the functions of the safety representatives should be carried out by the shop stewards, while others, such as the printing unions, favoured, wherever possible, a separation of the roles and functions. The former group of unions tended to argue that only the experience and 'power' of the shop stewards would give sufficient 'strength' to the safety representatives role, while the latter group of unions tended to be concerned about possible work overload and role conflict problems. In view of the extensive discussion of this issue we specifically consider, in the following chapter, whether there were any significant differences between the attitudes, behaviour, and potential impact of safety representatives who were shop stewards and those who were not.

Union discussions also urged the need to move beyond consultation over health and safety matters, as formally stipulated in the safety representative and safety committee regulations, to a position where this subject area becomes one of negotiation between unions and management. To this end a number of union and labour movement guidance documents (41) have argued that unions should strongly emphasise the safety representative function, rather than seeking to exert their major influence through the medium of joint health and safety committees, on the grounds that the legal rights of inspection and information specifically accorded to the safety representatives are more likely to provide the basis for making health and safety matters the subject of genuine negotiations. This matter is taken up in detail in the following chapter; indeed the determinants and implications of this consultation/negotiation distinction provide the central focus of our analysis of the questionnaire returns from a sample of safety representatives.

NOTES

1. John Mendeloff, *Regulating Safety: An Economic and Political Analysis of Occupational Safety and Health Policy*, MIT Press, Cambridge, Mass., 1979, p.7.

2. Guido Calabresi, *The Costs of Accidents: A Legal and Economic Analysis*, Yale University Press, New Haven, 1970.

3. See, for example, Roy Lewis and Geoff Latta, "Compensation for Industrial Injury and Disease", *Journal of Social Policy*, Vol.4, No.1, January 1975, p.47-50.

4. *Report of the Committee on Safety and Health at Work, 1970-72*, Cmnd 5034, HMSO, London, 1972, p.143-7.

5. See, for example, Howard P Marvel, "Factory Legislation: A Reinterpretation of Early English Experience", *Journal of Law and Economics*, Vol.20, 1977.

6. Sidney Webb, preface to B L Hutchins and A Harrison, *A History of Factory Legislation*, Frank Cass and Co Ltd, London, Third Edition, 1966.

7. K W Wedderburn, *The Worker and the Law*, Pelican, Harmondsworth, Second Edition, 1971, p.239.

8. L Greenberg, "Does Government Enforcement Help Industrial Safety?", *Engineering*, September 1972.

9. Cmnd 5034, *op.cit.*, p.66.

10. See, for example, P W J Bartrip and P T Fenn, "The Administration of Safety: The Enforcement Policy of the Early Factory Inspectorate, 1844-1864", *Public Administration*, Vol.58, Spring 1980.

11. W G Carson, "White Collar Crime and the Enforcement of Factory Legislation", British Journal of Criminology, 1970.

12. See, for example, P B Beaumont, "The Limits of Inspection: A Study of the Workings of the Government Wages Inspectorate", Public Administration, Vol.57, Summer 1979.

13. Owen A Hartley, "Inspectorates in British Central Government", Public Administration, Vol.50, Winter 1972.

14. Cmnd 5034, ibid., p.7.

15. Cmnd 5034, Loc. cit.

16. This discussion is very much based on R W L Howells, "Worker Participation in Safety: The Development of Legal Rights", Industrial Law Journal, 1974.

17. Howells, op. cit., p.88.

18. Department of Employment Evidence to the Safety and Health at Work Committee (Robens), 1970-72, Vol.2, Selected Written Evidence, HMSO, London, p.235.

19. Although for an earlier motion on this matter see Annual Report of the Trades Union Congress, 1954, p.426-9.

20. Annual Report of the Trades Union Congress, 1964, p.422.

21. Annual Report of the Trades Union Congress, 1964, p.423.

22. Annual Report of the Trades Union Congress, 1965, p.418.

23. Quoted in Department of Employment Gazette, November 1967, p.885.

24. Cmnd 5034, ibid., p.21-3.

25. The following summary draws heavily on Brenda Barrett, "Safety Representatives, Industrial Relations and Hard Times", Industrial Law Journal, 1977.

26. Barrett, op.cit., p.175-6.

27. See Industrial Relations Review and Reports, May and July 1976.

28. See P B Beaumont, Safety Legislation: The Trade Union Response, Occasional Papers in Industrial Relations, Universities of Leeds and Nottingham, No.4, 1979, p.21.

29. Beaumont, Loc.cit.

30. See, for example, Industrial Relations Services, Worker Participation in Health and Safety: A Comparative Study of the Law and Practice in Four European Countries, London, August 1980.

31. H C Debs, Vol.919, November 19, 1976, Col. 797.

32. *The Times*, December 2, 1976.

33. Health and Safety Executive, *Health and Safety: Manufacturing and Service Industries, 1977*, HMSO, London, 1978, p.6.

34. W W Daniel and Elizabeth Stilgoe, *The Impact of Employment Protection Laws*, PSI, Vol.XLIV, No.577, June 1978, p.81-4.

35. This discussion is based on P B Beaumont, "Explaining Variation in the Enterprise Response to Industrial Relations Legislation: The Case of the Safety Representative Regulations", *Personnel Review*, 1981.

36. Beaumont, *op.cit.*, p.23.

37. *IDS Study No. 218*, May 1980.

38. Quoted in *Department of Employment Gazette*, April 1978, p.396.

39. See, for example, M J Pedler, "The Training Implications of the Shop Stewards Leadership Role", *Industrial Relations Journal*, Vol.5, No.1, Spring 1974; John Lover, "Shop Steward Training: Conflicting Objectives and Needs", *Industrial Relations Journal*, Vol.7, No.1, 1976.

40. For further details see *IDS Brief*, No.181, May, 1980.

41. See, for example, Michael Cunningham, *Safety Representatives: Shop Floor Organisation for Health and Safety*, Studies for Trade Unionists, WEA, London, March 1978.

Chapter Five

THE SAFETY REPRESENTATIVE FUNCTION

The industrial relations legislation of many countries frequently draws a distinction between subjects over which there is to be joint consultation, as opposed to those over which there is to be negotiation. The former are those where the basic aims of union(s) and management are held to be essentially similar, while the latter are those where there is held to be a fundamental divergence of interests between the two parties. The problem-solving orientation of joint consultative arrangements with the possibility of mutual gains to union and management (i.e. a varying sum pay off matrix) has been labelled "integrative bargaining" by Walton and McKersie,(1) while the issue orientation of the negotiation process has been labelled "distributive bargaining". In the consultation process, at least ideally, the parties seek out all information relevant to the issues, maximise the amount of information exchanged, and avoid coercive and threatening tactics, whereas in negotiations the parties limit the amount of communication and information, engage in bluffing, attempt to establish their commitment to given positions, and use various forms of coercive behaviour, such as warnings, promises and threats.(2) Finally, it should be noted that in the eyes of unions consultation is typically seen as a less satisfactory form of union-management interaction than negotiation in that the ultimate right of decision making on the subject in question rests solely with management.

However, whether this legislative and academic (literature) distinction has any real practical relevance or significance at the level of the individual union-management relationship has long been open to question. In this regard the late Allan Flanders, for example, observed that,(3)

What has to be queried is not whether consultation is necessary to provide adequate means of communication in modern industry, but whether it is a satisfactory substitute for bargaining. In practice the dividing line between the two methods is often blurred or non-existent. When union representatives claim the right to be consulted they are more often than not demanding an opportunity to negotiate should the need for it arise. Where purely workplace issues are at stake, which have no repercussions on a broader front, negotiation may not involve full time union officials or result in formal agreements. It may be conducted by shop stewards, possibly within a procedure for joint consultation. This helps to sustain the myth that the issues have not been settled by collective bargaining at all, and that management has maintained its final responsibility for the decisions taken.

The above sort of view would appear to call into question the practical utility of making any distinction between the concepts of consultation and negotiation, as opposed to using a single, all embracing term such as, for example, 'joint deliberation'. It is this matter which provides the analytical focus of this chapter where we consider the basis for, and implications of, such a distinction using the questionnaire responses of a sample of union appointed safety representatives. This particular body of sample data is, potentially at least, of considerable value in examining this issue. This is because the subject area that has undoubtedly been considered the most suitable for joint consultation, at least in Britain, has been workplace health and safety. The reason for this being the strongly held belief in a basic similarity of union-management aims in relation to this particular subject. This point was forcefully made by the Robens Committee in the following terms,(4)

> There is a greater natural identity of interest between 'the two sides' in relation to safety and health problems than in most other matters. There is no legitimate scope for bargaining on safety and health issues but much scope for constructive discussion, joint inspection and participation in working out solutions.

It is this view which is, as we have seen, reflected in the formal provisions of the Health and Safety at Work Act 1974, which provide for union appointed safety representatives to consult (and be consulted by) the employer on arrangements for the joint promotion of "measures to ensure the health and safety at work of employees, and in checking the effectiveness of such measures." However, this formal obligation to consult (only) may mean little in practice because, as one labour movement guidance document has stressed, "... the union representative who consults and is consulted by the employer is also a negotiator."(5) Indeed a number of union and labour movement guidance documents have _specifically_ argued that unions should emphasise the role of the safety representative over that of joint health and safety committees on the grounds that the former is more likely to develop a negotiating role than the latter.(6) The reasoning underlying this suggestion is the fact that "the law which applies to safety committees is very limited. All powers of inspection, disclosure and information, etc. are given to the safety representatives NOT to the safety committees."(7) The safety representative is certainly a potentially very powerful individual at the workplace, being in fact the only union appointed representative at the workplace in Britain with statutory backing for their function.

Some Hypotheses Suggested in the Existing Literature

There are a number of explanations and implications of this consultation-negotiation distinction apparent in the relevant body of literature. The first explanation of the distinction is the one that was mentioned in the previous section, namely that an essential similarity of union-management aims in relation to a particular subject makes it suitable for consultation, whereas dissimilar aims are likely to make for union-management negotiation over the subject. According to McCarthy this dichotomy appears to paradoxically assume that,(8)

> ... management should only agree to share responsibility on controversial and conflicting subjects, like wages; on non-controversial and common interest issues, like manning, it cannot do more than consult. So we reach a position in which it is suggested that agreements are only possible when the two sides are basically

opposed; when they are really united there cannot be any question of an agreement.

However, despite this apparent paradox, it is the similarity (dissimilarity) of union-management aims that is the most frequently alleged basis for the consultation-negotiation distinction. However, there is a second explanation of this distinction which tends to be implicit in a good deal of the trade union literature on the subject. This is the notion or belief that consultation is a lesser form of union-management interaction, at least from the union point of view, which is essentially forced on unions by a management determined to reserve to itself the ultimate right of decision making on as many subjects as possible. That is, the unions are constrained to accept this 'less valuable' form of interaction by a management strongly committed to the maintenance of management rights and prerogatives.

The logical implication of the first explanation of this distinction is that there is relatively little conflict potential in consultation over any particular subject compared to a situation of negotiation over that same subject. In other words, negotiation has the connotations of 'hard bargaining' where an essentially zero sum game situation leads one party to try to gain at the expense of the other by drawing on its various sources of power in order to raise the other party's costs of disagreeing with its demands or offers. In contrast, the purely problem solving orientation of both parties in consultation is held to be facilitated and shaped by the presence of trust, mutual respect and information sharing, rather than by reference to any use of bargaining power or manipulative tactics. In contrast, the implication of the second explanation, which was outlined above, is that unions will 'get less out' of consultation than out of negotiation. That is, a union representative will be able to take fewer (if any) initiatives and consequently be less active in a situation of consultation, with the result that less of the union's aims in relation to the particular subject area will be capable of being achieved in the consultation situation compared to that of negotiation.

As we have already seen, there have always been certain reservations expressed about just how meaningful this consultation-negotiation distinction is in practice at the level of individual union-management relationship. Moreover, there is now a gradually accumulating body of empirical evidence which seems to suggest that there are in fact good reasons for being somewhat uneasy about the nature of the basis (if any) of this particular distinction. In Britain, for example, there have been a number of empirical studies which have indicated that the composition of the employee side of joint consultative committees is frequently not too dissimilar from that of negotiation bodies at the same workplace,(9) a finding which suggests that in practice there may be rather more negotiation over a subject than is implied by the fact that the union-management committee is formally titled a 'consultative' one. Furthermore, in their empirical investigation of the factors that appeared to facilitate the four sub-processes of bargaining (which were conceptually distinguished by Walton and McKersie) Peterson and Tracy found that, (10)

> ... there seems to be less direct conflict between the tactics used in integrative bargaining and those used in distributive bargaining than predicted by theory. Perhaps the mixed nature of most bargaining keeps the majority of negotiators from applying all out distributive tactics. At any rate, strong bargaining power, constructive relationships, clear and specific statements of issues, as well as exploring them in a noncommittal fashion, seem to aid both distributive and integrative bargaining.

Similarly, Kochan, Dyer and Lipsky's recent study of the effectiveness of joint health and safety committees reported that "although the predominate pattern is one of problem solving, a good deal of bargaining or negotiating behaviour also characterises the interactions between unions and employers on safety and health issues."(11) Specifically, what they found was that problem-solving and negotiating strategies were complementary behaviours for union representatives, but substitutes for management representatives. That is, the more the union engaged in problem solving behaviour, the more it also engaged in negotiating behaviour (and vice versa), but more problem-solving behaviour on the part of management was associated with less negotiating behaviour (and vice versa). Moreover, this generalisation held true for the factors determining the parties choice of strategies.

In view of this accumulating evidence it is important for industrial relations scholars to systematically re-examine the basis (if any) of the distinction between consultation and negotiation, and to consider any implications that appear to follow from the observed existence (if any) of such a distinction. This is the task undertaken here using the results of a questionnaire administered to a sample of union appointed safety representatives in Britain. In contrast to the body of evidence reviewed above, which was drawn from studies of committees and teams of negotiators, we are considering here the position of an individual representative. This is an important point to emphasise as the findings for the operation of committees may not carry over directly to the situation of an individual representative and vice versa. In view of this possibility it is important that the sort of study undertaken here be replicated and extended using other bodies of data drawn from both individual representatives and groups of individuals who are part of various institutional arrangements.

The Basic Characteristics of the Sample

The analysis and findings to be presented here are based on the responses to a number of questions contained in a questionnaire which was administered to a sample of 225 safety representatives attending health and safety training courses at the Central College of Commerce in Glasgow during the year 1979. This 'class room' method of gathering information on the attitudes and activities of union and work-group representatives is one that has been employed with considerable success in both Britain and the United States.(12) There is no reason to think that our sample is, in terms of their characteristics, attitudes and activities, significantly different from other groups of safety representatives undergoing training at the same time throughout the country, although it may not be entirely representative of the full population of safety representatives in the country at large. This is because of the fact that our group was one of the earliest in the union health and safety training programme. The fact that our representatives came overwhelmingly from highly organised plants is obviously suggestive in this regard, although it should be noted that the extent of union organisation was one of the few variables to exhibit relatively little variance. Moreover, their early participation in the training programme was the very reason that we sought to sample this particular group of representatives -

i.e. to provide an early benchmark for later, more detailed studies of safety representatives to build on.

The total number in our group (N = 225) represented approximately a one-in-three sample of the full number of safety representatives who attended such courses at the College during the 12 months after the safety representative regulations became law in Britain in October 1978. The sample was very much a male dominated one, with only 13.3 per cent of our respondents being women. A skill breakdown of the sample revealed that some 47 per cent of the representatives were skilled workers, 31.7 per cent semi-skilled and 18.7 per cent non-manual workers. There was considerable age variation among our representatives, with 11.6 per cent being aged below 25, 32.9 per cent 25-34, 21.8 per cent 35-44, and 26.7 per cent in the age range 45-54. The representatives tended to be very much long serving employees with their company as only 10.7 per cent had been employed there for less than 2 years, 21.5 per cent had been employed there for between 2 and 5 years, 21.7 per cent for between 5 and 10 years, and fully 36.6 per cent for more than 10 years. One point of particular interest, in view of the considerable discussion of the matter within the trade union movement, was the division between safety representatives only and those persons who were both safety representatives and shop stewards. In our sample, 59.1 per cent were both safety representatives and shop stewards (of whom 79 per cent had been shop stewards for 2 years or more), with the remainder being safety representatives only. The respective sizes of these two groups were very much a function of the distribution of individual unions (with their differing policies on the matter) in the sample. The median sized workforce constituency of our representatives was 50 persons, a figure rather larger than that reported in the latest study of shop steward constituency size.(13)

The representatives came overwhelmingly from highly organised plants, where union membership among the manual worker grades was typically above 70 per cent. Although the majority of the representatives were employed at workplaces with more than 200 employees, there was a considerable range of industries represented in the sample. The major industry groups represented were shipbuilding and the timber and furniture industry (both 14 per cent), mechanical engineering (13 per cent), and printing and clothing (both 9 per cent). The vast majority

(72.3 per cent) reported that health and safety matters were at least of some degree of concern to employees at their workplace, while 67 per cent saw health and safety matters to be of a similar degree of concern to their management. When questioned about the "general quality" of the union-management relationship at their workplace fully 76 per cent of them described it as a reasonably cooperative, as opposed to hostile, one. This judgement did not, of course, mean the absence of industrial conflict, as 66.2 per cent said that there had been at least one instance of industrial action (more broadly defined than strikes) at their workplace during the preceding year. And just under 20 per cent said that there had been some form of industrial action over health and safety matters during this period of time.

The Findings for the Distinction: Basis and Implications

This sample of union appointed safety representatives was asked to consider which of the terms 'consultation' or 'negotiation' best described the basic nature of their representative function. In response fully 72 per cent of the sample stated that they saw themselves as basically consulting with management, with the remaining 28 per cent choosing the description of negotiation. When questioned about their reasons for choosing between the two terms the minority who described themselves as negotiators overwhelmingly tended to explain their choice in terms of 'it was the only way to get things done' or 'it was the result of the management's attitude'.

In order to try and explore the nature of the underpinnings to these responses in more detail we sought to identify any systematic influences on whether the representative described himself as negotiating, or consulting, with management. The behavioural models of negotiations, although often laboratory rather than fieldwork based, with their emphasis on a variety of structural, organisational, and interpersonal variables, such as the incompatibility of the goals of the parties and the degree of trust between the parties, were a major influence on the nature of this exercise.(14) In fact the individual variables that seemed at least potentially relevant to the task of trying to account for the observed variation between the consultation and negotiation description can be conveniently grouped under the following sub-headings:-(15)

(i) <u>Variation in the environment in which the safety representative operates</u>
The individual variables employed here were plant size, industry of employment (the distinction here being between the high accident rate, shipbuilding industry and the rest), the number of instances of industrial action at the workplace in the preceding year, the number of employees for whom the individual safety representative was responsible, and whether there had been any industrial action specifically over health and safety matters during the preceding year.

(ii) <u>Variation in the personal characteristics of the safety representatives</u>
The individual variables here were age, length of service with the company, manual or non-manual status, and whether the safety representative was also a shop steward.

(iii) <u>Variation in the relevant attitudes and opinions of the safety representatives</u>
The individual variables employed here were the safety representatives view of the overall quality of union-management relations at his workplace, his view as to the extent of management's concern to minimise union involvement in plant level decision making, his assessment of the similarity of union-management aims in the health and safety area, his assessment of the extent of concern of both employees and management with health and safety matters, and his view of the extent of union influence over health and safety matters at his workplace prior to the passage of the Health and Safety at Work Act 1974.

This list is a mixture of <u>structural</u> and <u>cognitive</u> influences, with most of the variables being entered in binary form. In view of our earlier discussion of the basis of the consultation-negotiation distinction the two variables of particular interest here are (i) the safety representatives assessment of the similarity of union-management aims in relation to workplace health and safety, and (ii) his assessment of the extent of management concern to limit the range of union influence at the workplace. Accordingly, in Table 5.1 below we present the basic correlation results obtained from our attempt to identify any systematic influences on the minority (majority) of our safety representatives who described themselves as negotiators (consultors).

Table 5.1: Correlation Coefficients Between Reporting Basic Function as Negotiation and Independent Variables

Subvector	Individual Variables	Correlation Coefficients
(i)	Plant size	+ 006974
	Industry (Shipbuilding = 1)	- 0.04295
	Number of Instances of Industrial Action	+ 0.03644
	Size of Workforce Constituency	+ 0.11316
	Industrial Action over Health and Safety (Occurred = 1)	+ 0.06882
(ii)	Age (≤ 44 = 1)	+ 0.03790
	Length of Service	+ 0.01376
	Non-Manual Status (= 1)	- 0.03098
	Shop Steward (= 1)	+ 0.15104*
(iii)	Assessment of overall quality of union-management relationship (Essentially co-operative = 1)	- 0.12678*
	Assessment of extent of management concern to minimise union influence in decision making (Relatively concerned = 1)	+ 0.19125*
	Assessment of similarity of union-management aims on health and safety (Essentially similar = 1)	- 0.17903*
	Assessment of extent of employee concern with health and safety matters (Relatively concerned = 1)	- 0.07527
	Assessment of extent of management concern with health and safety matters (Relatively concerned = 1)	- 0.14095*
	Assessment of extent of union influence on health and safety prior to 1974 Act (Relatively influential = 1)	- 0.19926*

* = Statistically significant

The contents of the above Table indicate that there were six statistically significant influences on this dichotomy which were as follows: whether the safety representative was also a shop steward, the safety representatives view of the overall quality of the union-management relationship, his assessment of management's concern to minimise the range of union influence, his assessment of the similarity of union-management aims in relation to health and safety matters, his assessment of the extent of management concern with health and safety matters, and his assessment of the extent of union influence on health and safety at this workplace prior to the 1974 Act. The cognitive variables were clearly of far more influence in shaping the nature of this response than the structural variables. It is particularly important to note how the safety representatives attitudes seem to be especially strongly shaped by their perceptions of management attitudes to health and safety, union involvement in health and safety prior to the 1974 Act, and management attitudes to union influence more generally. (This is essentially the case with all the findings presented in this chapter). The importance of management attitudes, or more accurately in this case, the union perceptions of those attitudes, is in keeping with the findings of a number of well known studies, particularly the famous Illini City studies of the 1950s which concluded that the attitudes and policies of management were the major determinants of the overall climate of the union-management relationship in a plant.(16) The two variables of central interest to us here were both statistically significant, but one of them entered with an unexpected sign. This was the extent of management concern to minimise union influence variable, which indicated that safety representatives who felt that their management was relatively concerned to minimise the extent of union involvement in plant decision making would describe their basic function as one of negotiation. This finding was the opposite of the oft-heard suggestion that management's desire to limit union involvement in plant decision making would be a powerful constraint working to limit the safety representatives' definition of their function to that of consultation. In contrast, it would appear that where safety representatives felt that their management was not particularly interested in maintaining 'management rights' then consultation was sufficient to achieve their aims and objectives in the area of workplace health and safety. Conversely, where management was seen or felt to be generally strong on management rights then the safety representative reacted to their belief in this regard by adopting a negotiating posture on the grounds that it was necessary to proceed in this way in order to have any hope of achieving their desired health and safety aims and objectives. This finding seems consistent with the Flanders suggestion, which we noted earlier, that a negotiating stance or strategy will be adopted *if the need for it arises*, with the need likely to be the result of management's general attitude to the extent of union influence in decision making. The direction of this relationship is also consistent with Abell's more specific suggestion that,(17)

> ... managers will not only be the most important source of information for safety representatives, they will also influence the usage of information in the way that they provide and respond to it. For example, if they only provide it reluctantly and partially union representatives will tend to respond in terms of distributive bargaining tactics.

The interesting implication that would appear to follow from our finding for this particular variable is that in view of the widely accepted connotations of the term 'negotiation', namely hard bargaining and conflict potential, it would appear that a rigid insistence on management rights can be counter-productive from the management point of view, at least in terms of the attitudes and behaviour of union representatives that it generates in response.(18)

The signs on the other significant variables were all reasonably predictable and straightforward in their interpretation. A belief in the basic dissimilarity of union-management aims in relation to health and safety matters was positively associated, as predicted, with the safety representative describing his function as basically a negotiating one. The safety representatives who saw their management as being relatively unconcerned with health and safety matters were also most likely to describe themselves as negotiators; a finding whose interpretation parallels that for a management concerned to minimise the range of union influence in workplace decision making. In addition we found that safety representatives who reported that there had been relatively little union influence over health and safety matters at their workplace prior to the 1974 Act were more likely to describe their basic function as one of negotiation. The interpretation of this finding, which we would favour, is that in situations where management had been unwilling to concede union involvement in the subject area on a voluntary basis, the safety representatives have reacted to this sort of management attitude and position, which is still likely to have implications for their current operation, by adopting distributive bargaining attitudes in the belief that this is the only way in which they can hope to achieve their aims. Finally, we found that the safety representatives who were also shop stewards were much more likely to describe their function as a negotiating one. In this case it seems likely that these individuals believe that their general experience as stewards, together with the possibility of exerting pressure in other subject areas if their health and safety demands are felt to be thwarted by management, gives them much greater potential for adopting negotiating-like behaviour in performing their safety representative function.(19)

In a preliminary study, such as this, of such an important issue correlation analysis would appear to be more appropriate than multiple regression analysis. However, we did estimate a number of stepwise regression equations,(20) with the best fit equation (defined in terms of the step at which the adjusted R^2 was maximised) being as follows:

$$\text{Neg} = \underset{(t\,=\,2.2)}{0.928\ \text{CONST}} + \underset{(t\,=\,2.3)}{0.150\ \text{PACT}} + \underset{(t\,=\,2.1)}{0.143\ \text{MP}} + \underset{(t\,=\,2.3)}{0.142\ \text{SS}} - \underset{(t\,=\,1.8)}{0.120\text{SA}}$$

$(\bar{R}^2 = 0.09)$

The four variables in the best fit equation were, in order of entry, the safety representative's assessment of the extent of union involvement in health and safety prior to the 1974 Act (PACT), his assessment of the extent of management's concern to minimise the range of union influence (MP), whether the safety representative was also a shop steward (SS), and finally their view of the similarity of union-management aims in relation to health and safety (SA). The first three variables were statistically significant, but the similarity of aims variable fell just short of statistical significance which would suggest that it might have rather less influence in providing the basis for the consultation-negotiation dichotomy than tends to be suggested in the existing literature. This is certainly a finding that should be re-examined on the basis of other bodies of data.

As indicated earlier, the term 'negotiation'

has the connotations of hard bargaining, with substantial conflict potential, in which there is the expectation that the various sources of bargaining power will have to be drawn upon in order to bring about a settlement. If this is a valid implication of the consultation-negotiation distinction, at least in the minds of our individual safety representatives, then the distinction should influence the strategy and tactics which they employ in performing their safety representative function. On this particular matter we were able to obtain information from the responses to a question concerning their view of the single most important factor necessary for them to function effectively as a safety representative. The distribution of answers to this question was as follows:

Must be well informed about relevant safety regulations	54.3
Must be well informed about relevant causes of accidents	8.0
Must have full backing of workforce in activities	20.9
Must have ability to convince management of need for change in workplace organisation and practice	16.7

The emphasis of the first two categories of response above is very much on the need for an effective safety representative to be a well informed, well trained person capable of carrying out a relatively specialist technical function. These respondents tended to see the safety representative as performing a very different function from that of any other workgroup or trade union representative at the workplace. In contrast, the third category of respondents stressed the key importance of the workforce as a back-up resource and were thus less concerned with the relatively technical nature of the safety representative function, tending to see themselves as essentially the same as other union and workgroup representatives who could best achieve their aims by being able to refer to the needs, wishes and commitment of their workforce constituencies. The final category of respondents could, at least at first glance, have fitted into either of the above two groupings, but the answers to related questions indicated that they very much belonged to the first group. That is, they saw their ability to convince management of the necessity for change as very much deriving from the fact that they were well informed, well trained individuals performing a relatively specialist, technical function. In their view management would respond to the presentation of evidence of hazards provided by themselves rather than to arguments based on what the workforce wanted in this regard.

Accordingly, we differentiated those representatives who emphasised the key importance of workforce backing from the rest and sought to account for any systematic influences on the pattern of these responses. To do this we used the same basic set of potential explanatory variables, but with the addition of the consultation-negotiation distinction as a key potential explanatory variable. The major hypothesis to be tested here is that a safety representative who basically describes himself as negotiating with management is likely to emphasise the key importance of workforce backing for effective functioning. This expectation follows from Benedictus's specific comment that "early in his appointment it may be uncertain to what extent his workmates will support a safety representative's initiatives - yet without such support his demands must lose effectiveness."(21) More generally, various behavioural studies have indicated the importance of constituency influence in producing distributive bargaining behaviour.(22)

The basic correlation results obtained are those set out in Table 5.2. This exercise produced only two statistically significant variables, but most importantly one of these was the safety representative's view as to whether their function was basically a negotiation or consultative one. In this case we found, as expected, that a safety representative who saw his basic function as one of negotiation (as opposed to consultation) was more likely to cite the key importance of workforce 'back up' for the effective carrying out of his function. This finding we interpret to mean that where health and safety changes recommended by the representative were seen as likely to be the subject of hard negotiation and bargaining with management, with the obvious potential for conflict between them, there was much more emphasis placed on the need to carry the workforce with them in this task. This interpretation is consistent with that for our second statistically significant variable which indicates that those safety representatives most likely to

Table 5.2: Correlation Coefficients Between Major Reliance on Workforce as a Key Resource for Effective Functioning and the Independent Variables

Subvector	Individual Variables	Correlation Coefficients
(i)	Plant size	+ 0.05202
	Industry (Shipbuilding = 1)	+ 0.01709
	Number of Instances of Industrial Action	+ 0.00303
	Size of Workforce Constituency	+ 0.00346
	Industrial Action over Health and Safety (Occurred = 1)	- 0.06040
(ii)	Age (≤ 44 = 1)	- 0.00288
	Length of Service	+ 0.03775
	Non-Manual Status (= 1)	+ 0.01062
	Shop Steward (= 1)	+ 0.02708
(iii)	Assessment of the overall quality of union-management relationship (Essentially co-operative = 1)	- 0.04403
	Assessment of extent of management concern to minimise union influence in decision making (relatively concerned = 1)	+ 0.11518
	Assessment of similarity of union-management aims on health and safety (Essentially similar = 1)	- 0.08504
	Assessment of extent of employee concern with health and safety matters (Relatively concerned = 1)	+ 0.07373
	Assessment of extent of management concern with health and safety matters (Relatively concerned = 1)	- 0.19926*
	Assessment of extent of union influence on health and safety prior to 1974 Act (Relatively influential = 1)	- 0.10456
	Negotiation (=1)	+ 0.15709*

* = statistically significant

emphasise the importance of workforce backing saw their management as being relatively unconcerned about health and safety matters. Again we would argue that where the safety representatives anticipated difficulty in convincing management of the need to change workplace organisation and practice in the interests of health and safety, because of management's relative lack of concern with health and safety matters, they were more likely to value the workforce as a back up resource to this task of hard negotiation. Finally, there were two variables that fell just short of statistical significance. These were the safety representatives assessment of management's concern to minimise union influence in decision making, and his assessment of the extent of union influence on health and safety prior to the 1974 Act. The signs on these variables, together with their interpretations, are very much in line with those above. Again we estimated a series of stepwise regression equations which produced the following best fit (i.e. the step at which the R^2 adjusted for degrees of freedom was maximised) equation:

$$WFB = 0.356\,CONST - 0.188\,MCHS + 0.131\,NEG + 0.131\,ECHS - 0.108\,IAHS$$
$$(t = 0.9) \quad (t = 3.1) \quad (t = 2.1) \quad (t = 2.0) \quad (t = 1.5)$$

$(\bar{R}^2 = 0.07)$

The four variables in this best fit equation were, in order of entrance, the safety representative's assessment of the extent of management concern with health and safety matters (MCHS), whether the safety representative saw himself as basically a negotiator (NEG), his assessment of the extent of employee concern with health and safety matters (ECHS), and whether there had been any form of industrial action over health and safety matters (IAHS). The first three variables to enter the equation were all statistically significant, but the only one worth commenting on, in that it was not statistically significant in the correlation analysis, was the safety representative assessment of the extent of employee concern with health and safety matters. In this case we find that the safety representatives who emphasised the key importance of workforce backing were very much in situations where they felt that employees were relatively concerned about health and safety matters. The interpretation of this finding is that a safety representative was most likely to emphasise the importance of workforce backing in situations where he was reasonably confident that such backing would be forthcoming, due to the workforce's relatively high degree of concern with health and safety matters. This particular finding, which points to a factor conditioning the *ability* of the safety representative to draw on the workforce, nicely complements the other two significant variables, which point to circumstances influencing the *need* of the representative to draw on the backing of the workforce.

Finally, we come to the implication, which derives very much from the trade union view of the basis of the consultation-negotiation distinction, that consultation is a less useful and valuable form of union-management interaction to them in that it carries the connotations of a less active representative or one who is able to take relatively few initiatives. This is held to result from the constraining attitudes and behaviour of management, which insist that in a situation of consultation the ultimate right of decision making on the subject is reserved to management. As a consequence, it is argued that trade unions will achieve far less out of a consultation situation than out of one of negotiation.

In order to provide some perspective on this matter we report the varying responses to (and factors associated with) the following two questions: (i) the safety representative's assessment of the interest of the workforce in his function and (ii) the safety representative's assessment of his impact on management's interest in health and safety matters. Specifically, we will be concerned to see whether those safety representatives who described themselves as basically negotiators attracted, at least in their view, relatively more workforce interest and had relatively more impact on management's attitudes towards health and safety. These two effects or impacts are certainly consistent with the notion that the negotiator is more active and thus capable of bringing about more change than his counterpart who describes himself as consulting only with management. However, in relation to the first question it must be assumed, for this effect to come about, that the safety representative's description of himself as a negotiator has certain attitudinal or behavioural characteristics that are obvious to the workforce constituency, which is surely not an unreasonable assumption if the distinction is to have any great operational significance.

The responses to the first question were as follows:

Strongly interested:	14.3 per cent
Somewhat interested:	50.7 per cent
Little interested:	27.8 per cent
Completely uninterested:	7.2 per cent

If we aggregate the first two categories then we find that just under two-thirds of the representatives felt that the workforce was relatively interested in their activities as safety representatives. In response to the second question, which asked the safety representatives to assess their impact on management's interest in health and safety matters, we obtained the following distribution of answers:

Very strong impact:	10.0 per cent
Strong impact:	26.7 per cent
Some impact:	49.3 per cent
Little impact:	9.5 per cent
No impact:	4.5 per cent

Although we are considering here only the short run impact of their appointment and activities it is interesting that only a little over a third (36.7 per cent) of the representatives considered that they had had a relatively strong impact on the priority that their managements accorded to health and safety matters. Perhaps predictably, just under half of the sample opted for the 'some impact' category. Accordingly, in what follows we seek to identify any systematic influences on those safety representatives who, respectively, reported that their workforce were relatively interested in their work as safety representatives, and that they had had a relatively strong impact on managements' attitudes to health and safety matters. Once again we use the same three sub-vectors of potential explanatory variables, with particular interest attracting to the statistical significance (or not) of the negotiation variable. The basic correlation results which were obtained for question (i) and (ii) respectively are set in Table 5.3 overleaf.

The negotiation variable is negatively signed in both cases indicating a relatively lesser impact of safety representatives who basically describe themselves as negotiators, but too much should not be made of the direction of the relationship in view of the fact that both coefficients fell well short

Table 5.3: Correlation Coefficients Between Responses to Impact Questions and the Independent Variables

Subvector	Independent Variables	Correlation Coefficients (i)	Correlation Coefficients (ii)
(i)	Plant Size	+ 0.06744	- 0.07686
	Industry (Shipbuilding = 1)	+ 0.09146	- 0.02147
	Number of Instances of Industrial Action	+ 0.00232	+ 0.01223
	Size of Workforce Constituency	- 0.01660	+ 0.01845
	Industrial Action over Health and Safety (Occurred = 1)	+ 0.01661	- 0.01821
(ii)	Age (≦ 44 = 1)	- 0.00464	- 0.11450
	Length of Service	- 0.04678	+ 0.07720
	Non-Manual Status (= 1)	- 0.13127*	+ 0.03760
	Shop Steward (= 1)	+ 0.07298	- 0.05151
(iii)	Assessment of the overall quality of union-management relationship (Essentially co-operative = 1)	+ 0.02441	+ 0.19526*
	Assessment of extent of management concern to minimise union influence in decision making (Relatively concerned = 1)	+ 0.00203	- 0.17441*
	Assessment of similarity of union-management aims on health and safety (Essentially similar = 1)	+ 0.07863	+ 0.14675*
	Assessment of extent of employee concern with health and safety matters (Relatively concerned = 1)	+ 0.32002*	+ 0.16261*

Assessment of extent of management concern with health and safety matters (Relatively concerned = 1)	+ 0.18934*	+ 0.15086*
Assessment of extent of union influence on health and safety prior to 1974 Act (Relatively influential = 1)	+ 0.02383	+ 0.15661*
Negotiation (= 1)	- 0.09474	- 0.09749

* = statistically significant

of statistical significance. The strength of this influence was, in both cases, very much less than that of a number of other factors. In this regard one notes that there was relatively more workforce interest in the safety representative where the representative was a manual employee, and would therefore presumably be responsible for solely manual employees, and in situations where both employees and management were seen to be relatively concerned with health and safety matters. And there was relatively more impact on management's attitude to health and safety matters where, in addition to both employees and management being relatively concerned about health and safety matters, there was an overall cooperative union-management relationship, the management had relatively little interest in trying to restrict the range of union influence in decision making, and there had been considerable union involvement in health and safety matters prior to the passage of the 1974 Act. The relatively slight influence of the negotiation factor was shown by the following best fit regression estimates:

$$\text{(i)}\quad WFI = \underset{(t\,=\,0.3)}{0.120\,CONST} + \underset{(t\,=\,4.3)}{0.311\,ECHS} + \underset{(t\,=\,1.7)}{0.114\,MCHS} - \underset{(t\,=\,1.8)}{0.144\,NM} + \underset{(t\,=\,1.4)}{0.93\,SS} - \underset{(t\,=\,1.1)}{0.080\,NEG} \qquad (\bar{R}^2 = 0.12)$$

$$\text{(ii)}\quad IMA = \underset{(t\,=\,1.1)}{0.511\,CONST} + \underset{(t\,=\,2.3)}{0.183\,UMR} + \underset{(t\,=\,1.9)}{0.135\,PACT} + \underset{(t\,=\,1.7)}{0.130\,ECHS} - \underset{(t\,=\,1.5)}{0.118\,MP} \qquad (\bar{R}^2 = 0.07)$$

The negotiation variable admittedly enters the first best fit regression, but with a negative sign which is nowhere near statistical significance. Indeed, only employee concern with health and safety matters is statistically significant here, although management concern with health and safety and non-manual status fall only just short of statistical significance and are certainly more powerful than the negotiation influence, as is shop steward status. The negotiation variable does not even enter the best fit regression on management impact where only the overall quality of the union-management relationship is significant, although union involvement prior to the 1974 Act and employee concern with health and safety only just fail to achieve this status.

Conclusions

In conclusion, the findings of this chapter may be summarised as follows:

(1) There were certainly found to be systematic influences on whether the safety representative described himself as negotiating, as opposed to consulting, with management. However, one of the major influences on this response (i.e. the management prerogative finding) was not entirely in the predicted direction.

(2) The safety representative's description of himself as a negotiator was systematically related to an important component of the strategy he adopted in carrying out his function, namely the emphasis on having workforce backing to function effectively.

(3) The safety representative's description of himself as a negotiator was not significantly related to the two self assessed measures of impact. These findings must therefore cast some doubt on any simple assumption or belief that a negotiator will necessarily achieve more, from the union point of view, than a representative who describes himself as consulting with management.

These preliminary findings would seem to suggest that there is certainly some basis and substance to the consultation-negotiation distinction, at least as reflected in the attitudes of the individual union representative concerned. However, a great deal more research is necessary on a much wider variety of data sources pertaining to both individual representatives, as well as teams of individuals, before we can be even remotely confident that we comprehend both the basis for and implications of this much discussed distinction in the industrial relations literature.

On the specific role of safety representatives our findings pertain only to the short run impact of the safety representative function, which raises the obvious question about its longer term future development and impact. In speculating about the future one might want to argue that with all the attendant publicity and interest that surrounded the safety representative regulations becoming law in October 1978 there was inevitably bound to be a considerable short run impact from the presence of safety representatives. However, as things settle down, it is possible that the initial enthusiasm and interest that they have generated will be very much reduced. On the other hand, one could argue that their influence and impact will tend to increase substantially with the passage of time. The reason for this being that at the present time safety representatives are only feeling their way in a relatively new subject area, but as they acquire more training and experience the confidence in their ability to handle the various aspects of their function will increase and so one can look forward to more momentum, activity and impact through time. These differing possible lines of development suggest that it would be of considerable value to take a sample of safety representatives and monitor their performance over a period of time with a view to observing any significant changes in their attitudes and behaviour. In the absence of this type of exercise it is to be hoped that the results presented in 'early days' studies, like that undertaken here, will provide a useful benchmark for subsequent studies to develop further.

NOTES

1. R E Walton and R B McKersie, A Behavioural Theory of Labour Negotiations, McGraw Hill, New York 1965

2. This description is very much based on Thomas A Kochan, Lee Dyer and David B Lipsky, The Effectiveness of Union-Management Safety and Health Committees, The W E Upjohn Institute for Employment Research, Kalamazoo, Michigan, 1977, p.46

3. Allan Flanders, The Fawley Productivity Agreements, Faber, London, 1964, p.241-2

4. Report of the Safety and Health at Work Committee, 1970-72 (Robens), Cmnd 5034, HMSO, London 1972, para. 66, p.21

5. LRD Guide for Safety Representatives, Labour Research Department, London, 1978, p.8. The most useful discussion of this issue is probably Roger Benedictus, Law at Work: Safety Representatives, Sweet and Maxwell, London, 1980, p.28-31

6. See, for example, Michael Cunningham, Safety Representatives: Shop Floor Organisation for Health and Safety, Studies for Trade Unionists, WEA, London, March 1978

7. ASTMS, Guide to Health and Safety at Work, London, undated, p.41

8. W E J McCarthy, The Role of Shop Stewards

in British Industrial Relations, Research Paper No.1 Royal Commission on Trade Unions and Employers Associations, HMSO, London, 1966, p.36

9. See, for example, A I Marsh et al, Workplace Industrial Relations in Engineering, Kogan Page, London, 1971, p.52

10. Richard B Peterson and Lane Tracy, "Testing a Behavioural Theory Model of Labor Negotiations", Industrial Relations, Vol.16, No.1, February 1977, p.50

11. Kochan, Dyer and Lipsky, op.cit., p.45-59

12. See, for example, William Brown, Robert Ebsworth and Michael Terry, "Factors Shaping Shop Steward Organisation in Britain", British Journal of Industrial Relations, Vol.XVI, No.2, July 1978; Thomas A Kochan, David B Lipsky and Lee Dyer, "Collective Bargaining and the Quality of Work: The Views of Local Union Activists", Proceedings of the Industrial Relations Research Association, Winter, 1974

13. Brown, Ebsworth and Terry, op.cit., p.141

14. For two recent reviews of these models see John M Magenau and Dean G Pruitt, "The Social Psychology of Bargaining: A Theoretical Synthesis" in G M Stephenson and C J Brotherton, Industrial Relations: A Social Psychological Approach, Wiley, London, 1978; and W Clay Hamner, "The Influence of Structural, Individual and Strategic Differences in Bargaining Outcomes: A Review", in D L Harnett and L L Cummings, Bargaining and Personality: An International Perspective, Dame Publications, Houston, 1979

15. In terms of fieldwork based studies the framework of analysis utilised here is most similar to that in L Dyer, D P Lipsky and T A Kochan, "Union Attitudes Towards Management Co-operation", Industrial Relations, Vol.16, No.2, May 1977

16. See Milton Derber, W E Chalmers and Milton Edelman, Plant Union Management Relations, University of Illinois Press, Urbana, Illinois, 1965

17. D Abell, "Industrial Relations and Safety Representation", Personnel Review, Vol.8, No.3, Summer, 1979, p.32

18. In this regard see Robert Dubin, "Attachment to Work and Union Militancy", Industrial Relations, Vol.13, No.1, February 1973

19. For an excellent case study illustration of this particular attitude and position see the interview with the senior safety representative at Rolls Royce, Hillington, which is reported in Scottish Trade Union Review, No.1, Spring 1978, p.13-15

20. It is acknowledged that the use of ordinary least squares to estimate an equation containing a dichotomous dependent variable may give rise to certain statistical biases. Theoretically more sound techniques do exist (e.g. logit and probit analysis), but studies employing these and ordinary least squares have in practice found little difference between the two sets of results. See, for example, Morley Gunderson "Retention of Trainees: A Study with a Dichotomous Dependent Variable", Journal of Econometrics, Vol.2, April 1974

21. Benedictus, *op.cit.*, p.31

22. Magenau and Pruitt in Stephenson and Brotherton (eds.), *op.cit.*,p.192-3. Given the findings of these studies there is arguably a two-way relationship between workforce constituency pressure and the presence of negotiating attitudes and behaviour.

Chapter Six

THE JOINT HEALTH AND SAFETY COMMITTEE FUNCTION

The relevant regulations of the Health and Safety at Work Act 1974 provide that when at least two safety representatives request in writing the establishment of a joint health and safety committee the employer must set up such a committee within a period of three months. The nature of the relationship between the safety representative function and the safety committee function is an interesting if, as yet, somewhat unclear one. The Robens Committee Report, for example, had this to say on the matter,(1)

> The proposal in the original government Bill was for the appointment of employees safety representatives who could then, if they wished, require the establishment of joint safety committees. Although the two elements of the proposal are closely related, they can be considered separately. We are in no doubt that the concept of employees safety representatives is more important than the concept of joint safety committees. Whilst there may be some substance in the argument that joint safety committees can only be effective if the desire to have them and make them work exists on both sides, we are not convinced that this argument has precisely the same application to employees safety representatives. The question is much less one of philosophy than of what would actually happen in practice.

The above perspective has been taken a stage further by a number of individual union and labour movement guidance documents which, as indicated in earlier chapters, have argued that the unions should very much look to the safety representative, rather than the joint health and safety committee, function as the channel through which to exert influence in seeking to improve workplace health and safety conditions. An ASTMS policy document specifically made the following points about joint health and safety committees:(2)

i A committee is not the best method of dealing with urgent issues.

ii A committee must have decision taking power, including the right to spend money.

iii No issue which any union is taking through procedure, or is in dispute about, should be discussed by the committee.

iv No safety representative can delegate or give up any of his rights to a committee.

v Whether or not there is a joint committee with management, union safety representatives should themselves have a committee.

There has been other guidance literature which has also emphasised the potential value of forming a safety representatives committee to take matters through the normal grievance procedure, or through a special health and safety grievance procedure, rather than through a joint health and safety committee.(3) The underlying belief here is that the former structure is more likely to result in health and safety matters becoming the subject(s) of negotiation, rather than consultation, than would be the case with the latter arrangement. This view may be somewhat exaggerated if safety representatives, especially those who are also shop stewards, are well represented on the employee sides of joint safety committees.(4) In this regard it is important to note that safety representatives have the power to have 'unsatisfactory' safety committees reconstituted.(5)

In contrast to the historical reluctance of many employers to establish joint health and safety committees on a voluntary basis there is recent evidence to suggest that the relevant regulations of the Health and Safety at Work Act have already led to a significant increase in the establishment of such committees. This increased establishment of committees will be examined in the following section. The regulations also appear to have led to some changes in the composition and functions of already existing committees, an influence that is also briefly examined here. Following a review of these matters we move on to develop a framework of analysis, based on the results of a number of relevant studies in the United States, that appears to have considerable potential for assessing the effectiveness (or not) of such committees. This discussion of the measurement, and determinants, of committee effectiveness should be seen against the background of some of the comments made in Chapter 2 where we considered the question of measuring the health and safety performance of the industry or establishment.

The Establishment of Joint Health and Safety Committees

The setting up of a joint health and safety committee is likely to raise a number of specific sources of potential disagreement between unions and management that will have to be resolved through a process of negotiation. In this regard the following issues or areas are likely to be particularly prominent:(6) the relationship of the committee to the normal negotiation process; the representation of non-union employees; the question of 'agreed' committee objectives; the rights of the committee in relation to those of the safety representatives; and the level of management, in terms of the extent of decision making authority within the management hierarchy, that is represented on the committee. The particular influence of some of these factors in holding back the establishment of joint health and safety committees can be seen in certain industries. In the national health service, for example, strong disagreement over whether the non-TUC affiliated, professional associations should be represented on such committees has been a major stumbling block to their establishment. More generally, however, it would appear that such potential problems have been resolved in view of the evidence of a quite substantial increase in the number of joint health and safety committees established since the publication of the Health and Safety at Work Bill. The evidence that we draw on in this regard is provided by the SSRC Industrial Relations Research Unit's survey of workplace industrial relations, which was previously utilised in some of the analysis presented in Chapter 3. The relevant evidence provided by this survey is set out in Table 6.1.

Although in total 82.1 per cent of the 970 sample establishments had a joint health and safety committee at the time of the survey in December 1977

Table 6.1: Joint Health and Safety Committees by Industry (%)

Industry	Health and Safety Committees*			
	(1) Voluntarily Established	(2) Established since 1974 Bill	(3) No Committees	(4) Total Establishments
Food	40.0	43.6	15.5	110
Coal Products	40.0	60.0	-	5
Chemicals	50.7	37.7	11.6	69
Metal Manufacture	59.2	27.6	10.5	76
Mechanical Engineering	45.8	34.4	18.8	96
Instrument Engineering	52.9	29.4	17.7	17
Electrical Engineering	38.5	43.8	14.6	96
Shipbuilding	81.8	-	18.2	11
Vehicles	53.7	32.8	13.4	67
Metal goods not elsewhere specified	39.0	42.1	19.0	95
Textiles	53.0	28.8	18.2	66
Leather	18.2	54.6	27.3	11
Clothing	17.9	47.8	32.8	67
Bricks	44.0	44.0	12.0	25
Timber	40.6	28.1	31.3	32
Paper	47.1	29.9	23.0	87
Other manufacturing	47.5	32.5	17.5	40
Total	44.4	36.7	17.9	970

*There were 9 committees whose age was reported as unknown. This accounts for the fact that in some industries the sum of columns (1) - (3) is less than 100%

- January 1978, fully 45 per cent of these had been set up since the publication of the Health and Safety at Work Bill. If we consider only those industries with a reasonable number of sample observations (i.e. 20) we find considerable variation in the extent to which establishments have set up joint health and safety committees under the 'shadow of legislation', ranging from 44 per cent in the Bricks industry to 27.6 per cent in metal manufacture. It is this sort of variation that poses the basic question that we are concerned with here, namely, why did establishment A move quickly to set up a joint health and safety committee following the publication of the Health and Safety at Work Bill, and not establishment B?(7)

In order to try and answer the above question we need to identify the relevant characteristics of establishments in column (2) of Table 6.1 that significantly differentiate them from those in column (3). This test was undertaken by making use of the model developed in Chapter 3 which sought to identify the relevant characteristics of establishments that had set up such committees on a voluntary basis. This model consisted of the following three sub-vectors of variables:

1. Variation in Workplace Health and Safety Hazards The individual variables employed here were the mean industry accident rate (at the MLH or 3 digit industry level) and three technology dummies (continuous process; large batch and mass production; small batch).

2. Variation in Union Power: The individual variables employed here were union density, multi-unionism, shop steward organisation and the extent of industrial conflict.

3. Variation in the Personnel Orientation of Management: The individual variables employed here were establishment size, the existence of a general purpose consultative committee, foreign ownership and the presence of a member of senior management with specialist responsibility for industrial relations and personnel matters.

These thirteen variables were entered stepwise in a discriminant function with three of them being positively and significantly associated with the recent establishment of a joint health and safety committee; namely, the continuous process technology

dummy, the industrial relations management variable and foreign ownership. As the latter two variables are of considerable potential interest in telling us something about the characteristics of establishments that appear to be particularly sensitive, at least in terms of their speed of response, to the needs and demands of industrial relations legislation in general they warrant some further discussion. The first point to note in considering the interpretation and implications of our findings is that there has been considerable discussion of the effects of the previous Labour government's programme of industrial relations legislation in raising the status of the personnel-industrial relations management function within the management hierarchy of many organisations during the 1970s.(8) Against the background of this general discussion we can place our first finding which seems to indicate that where such relatively high status has been obtained, i.e. representation at the most senior level of management decision-making within the establishment - this has, in turn, the effect of making the establishments management especially conscious of, and responsive to, the provisions of industrial relations legislation. The result is that these establishments are disproportionately represented amongst the sub-group of firms that move quickly to meet the provisions of such legislation. The significance of the foreign ownership variable seems to provide some useful perspective on the oft heard claim that foreign owned firms, being strongly influenced by their parent companies' industrial relations policies and practices, tend to disregard many of the industrial relations conventions and practices that have long characterised the British system of industrial relations; a general reluctance to join employers associations and engage in multi-employer, industry based bargaining being a much publicised example in this regard.(9) In addition to the contents of Table 6.1, the survey conducted by the Health and Safety Executive in October 1979, some of the results of which were reported in Chapter 4, provided a number of useful findings with regard to joint health and safety committees. This survey sought to assess the impact of the regulations which came into force in October 1978 by considering the extent to which establishments where safety representatives had been appointed also had joint health and safety committees. In total they found that 81% of the workplaces where safety representatives had been appointed also had joint health and safety

Table 6.2: Joint Health and Safety Committees in Workplaces with Safety Representatives, by Industry

Industry	Joint Health and Safety Committees (i) Committee Resulted from regulations	(ii) Committee altered as a result of regulations
Agriculture, forestry	70.0	30.0
Mining and quarrying	51.0	12.0
Food, drink and tobacco	29.0	29.0
Coal and petroleum products	22.0	11.0
Chemical and allied	13.0	38.0
Metal manufacture	19.0	23.0
Mechanical engineering	32.0	18.0
Instrument engineering	0.0	50.0
Electrical engineering	40.0	23.0
Shipbuilding	17.0	33.0
Vehicles	29.0	17.0
Metal goods not elsewhere specified	39.0	19.0
Textiles	18.0	21.0
Leather, leather goods, fur	0.0	0.0
Clothing and footwear	50.0	20.0
Bricks, pottery, glass and cement	30.0	26.0
Timber, furniture	53.0	27.0
Paper, printing and publishing	29.0	27.0
Other manufacturing	30.0	24.0
Construction	59.0	14.0
Gas, electricity and water	54.0	9.0
Transport and communications	33.0	6.0
Distributive trades	35.0	30.0
Insurance, banking and business services	0.0	0.0
Professional and scientific services	58.0	24.0
Miscellaneous services	48.0	23.0
Public administration	56.0	24.0

Source: Department of Employment Gazette, February 1981, p.57

committees. The contents of Table 6.2 indicate the industry distribution of workplaces where (i) committees had been established as a result of the regulations and (ii) committees had been altered as a result of the regulations. The new committees tended to be established in the smaller plants and in the lower accident rate industries, which is where voluntarily established committees had been significantly under-represented. The survey reported that the constitution or structure of voluntarily established committees had been altered as a result of the regulations, although no details were provided with regard to the nature of these alterations.

In order to provide some feel for the nature of the changes and reforms that have occurred to certain committees as a result of the regulations, we turn to a number of case studies that have been conducted. The first such case study is the Vickers South Marston Works, an engineering establishment whose main product is hydraulic valves and pumps and were there has been a joint health and safety committee since 1974.(10) The composition of this safety committee, which met on a monthly basis, was as follows: seven safety representatives for staff employees, the safety manager (Chairman of the Committee), the safety officer,(Secretary of the Committee), and the site engineer. In view of the forthcoming regulations, the unions and management began discussions on possible revisions to the safety committee structure early in 1976. The particular focus of these discussions were on the need to more fully involve the non-manual unions as well as to ensure that all employees throughout the site, and not simply those in the hydraulics and services sections, were adequately represented on the committee. The initial management proposals were that there should be 4 safety representatives from the manual unions, 4 from the staff unions, with the personnel manager to be Chairman of the Committee, the safety officer to be Secretary, and with the safety manager and site engineer also represented. The union response was a proposal for an 18 member committee, with 6 members from the manual unions, 6 from the non-manuals and 6 management representatives, with the Chairman and Secretary of the Committee to be appointed on an annual basis by the members of the Committee. However, management felt that a committee of such size would be a too unwieldy body and therefore suggested the possibility of establishing three separate committees covering each of the main activities carried out on the site. These possibilities were still under discussion at the time the study was conducted.

Another case study worth noting is that of Molins Limited, Deptford, which primarily makes machinery for the tobacco industry. The study of this company reported that,(11)

> Molins is typical of many companies in the way it has taken account of the new legal provisions. Like many companies with a strong trade union presence, health and safety did not suddenly appear as a novel concern following the passage of the Health and Safety at Work Act; in fact only relatively small changes had to be made to the existing structures for health and safety.

The company had a joint health and safety committee in existence since the early sixties, but since the passage of the relevant regulations there have been major changes to the committee in terms of (i) the level of management representation and (ii) commitment to action which flows from the allocation of individual responsibilities. In the latter regard the chairman decides and minutes the allocation of specific actions to appropriate management members and arranges action himself on matters that fall outside the scope of these members. The committee, which meets monthly, consists of 20 members: 8 are management, including the production manager, the personnel manager, the commercial manager, and the training manager, with 12 safety representatives, elected at their respective union annual general meetings, constituting the remainder of the committee. The ability to generalise from any two case studies is always open to doubt, but it is worth noting the strong interdependency between the safety representative and safety committee functions in these two establishments. The significance of this fact can be gauged by recalling the concern expressed in a number of labour movement documents that management will try to 'water down' the potential power of the safety representative function by emphasising the importance of the safety committee function.

The strong representation of safety representatives on the employee side of joint safety committees is likely to have a number of important implications for the operation of such bodies, not the least of which is likely to be the potential for eliminating some of the more obvious weaknesses of earlier committees, such as lack of member training, and poor communication with the workforce at large, revealed in studies like that conducted by the Ministry of Labour in the late 1960s.(12)

The Effectiveness of Joint Health and Safety Committees

In principle at least joint health and safety committees can operate to improve workplace health and safety in two ways: (i) Directly, that is the suggestions and recommendations of the committee could lead to changes in workplace organisation and practice that have been identified as being largely responsible for hazards and accidents; and (ii) Indirectly, that is the committee's very existence could provide a focal point for the receipt of workers suggestions for improving workplace health and safety that would be acted upon, with the results being fed back to the workforce at large thus helping to raise the level of worker safety consciousness. In short, joint health and safety committees could hope to operate via these two routes, on at least some of the key work and worker characteristics that, as we saw in Chapter 2, have been held to be responsible for industrial accidents. However, even this schema suggests the likelihood of considerable variation in the potential effectiveness of joint health and safety committees in bringing about such improvements: for example, the above suggested that among the necessary prerequisites for such a favourable impact would be a management commitment to monitoring the sources of accidents; a willingness to implement the recommendations of the committee for often major changes in working practices and arrangements, often in the fact of conflicting production-technical considerations; and a workforce relatively concerned with health and safety matters and therefore interested in the work of the committee. The important point to make here is that even such basic pre-conditions for 'success' are unlikely to be randomly distributed across all establishments. Accordingly, the factors likely to account for such variation will be discussed at some length in the remainder of this Chapter.

The existing evidence on the impact of joint health and safety committees on industrial accidents, much less health and safety conditions in general, is far from adequate in nature. Certainly, reports of the Factory Inspectorate have, at various times, hailed the success of joint committees in this regard. As one such report put it,(13)

> During the last five years the incidence of notifiable accidents ... has declined in factories generally from 28 per 1,000 workers employed to 22.5, a reduction of about 20 per cent. This is in itself impressive evidence of increasing awareness on the part of both managements and operatives of their effective responsibilities in the promotion of safety. But what is perhaps most striking is that in undertakings where during this five year period active accident prevention organisations have been fostered substantially greater reductions have been achieved. In a chemical works, for example, employing 900 persons, the frequency rate has gone down from 4.0 to 2.41; in a motor vehicle works the rate has been cut by 50 per cent. These are not isolated examples ... but they do imply unmistakably that it is the firms with active accident prevention organisations which are really succeeding in eliminating avoidable accidents, and the fact that their success can be measured over a period of years shows that there is nothing fortuitous about their achievement.

It was no more than this sort of evidence that provided the basis for Williams' very strong conclusion that,(14)

> These examples establish as a fact that substantial reductions can be achieved in accident rates by safety organisation at the place of work; and, moreover, at a steady annual rate of reduction. But we know that only a small proportion of factories have established such organisations. What would be the position if all factories were similarly organised? It would mean that in the following year the total number of accidents could be reduced by at least 5 per cent; from, say, 180,000 to 171,000. And in subsequent years similar successive reductions would be likely.

Although Williams is not necessarily incorrect in this assertion, it is true to say that the present evidence is quite simply insufficient and inadequate to justify such a hard and fast conclusion at this stage. The fact of the matter is that it is difficult to demonstrate such a clear, causal relationship given the presently available sources of data. As an illustration of the difficulties involved in trying to identify a favourable impact of health and safety committees on the basis of existing data sources, we computed the Pearson correlation coefficient for the percentage of factories in the 16 2 digit industry groups reported by the Factory Inspectorate in 1969 to have a joint health and safety committee (see Chapter 3), and the relevant industry accident rates for the year 1970. The basic figures are those set out in Table 6.3.

Table 6.3: Accident Rates and Joint Health and Safety Committees, by Industry

	(1) Accident Rate (1970)	(2) Joint Health & Safety Committees (1969)
Food, drink and tobacco	38.2	32.0
Chemicals and allied	35.6	42.0
Metal manufacture	78.9	47.0
Mechanical engineering	38.6	28.0
Instrument engineering	15.1	28.0
Electrical engineering	24.6	28.0
Shipbuilding and marine engineering	72.4	34.0
Vehicles	34.3	35.0
Metal goods not elsewhere specified	39.7	26.0
Textiles	26.8	24.0
Leather and leather goods	18.4	19.0
Clothing and footwear	7.9	9.0
Bricks, pottery, glass and cement	54.2	27.0
Timber, furniture, etc.	31.3	13.0
Paper, printing and publishing	24.8	34.0
Other manufacturing	34.6	27.0

Sources: (1) Annual Report of H.M. Chief Inspector of Factories, 1970, Cmnd 4758, London, Appendix 7, p.101-3; (2) Department of Employment Gazette, July 1969.

Admittedly the resulting coefficient (0.633) was significant at the .01 level, but it was positively signed indicating that more committees are associated with a higher accident rate. The direction of this relationship was in fact reasonably predictable and as such provides a number of useful methodological lessons. First, if joint health and safety committees do have a favourable impact on industrial accidents this is only likely to be revealed by a close monitoring of their operations through time, although this will itself pose the obvious problem of adequately controlling for any other changing sources of influence on accident rates. An examination of this issue solely on a point of time or cross section basis is only likely to pick up the fact that high accident rate industries will, out of obvious concern for the problem, have established an above average number of joint health and safety committees. This was the case in our Chapter 3 analysis of the characteristics of plants having voluntarily established joint safety committees. Secondly, as we indicated in Chapter 2, there typically exists a substantial amount of underreporting of legally notifiable accidents. However, in establishments where there exist joint health and safety committees it is arguable that there may be a far more accurate reporting of accidents than elsewhere, thus giving the impression of even higher accident levels in well organised plants and industries than is truly the case. And finally, one should ideally enter as the independent variable a measure of the 'activity levels' of these committees, rather than simply the number which exist, as only active, well functioning committees, as opposed to ones that exist largely in name only, are likely to have any real chance of significantly reducing the level of industrial accidents. This latter point leads logically to the suggestion of a research approach that would involve devising measures or proxies of a successfully operating committee that are a necessary, if not sufficient, condition for a committee to have a favourable impact on industrial accidents. The second stage of the exercise would then involve seeking to identify the key environmental and organisational characteristics that are responsible for producing such an active committee.

In seeking to build up such an analytical framework we take as our starting point the work of Walton and McKersie (15) who identified four separate bargaining processes, each having quite distinct goals with their own underlying determinants. The particular sub-process of collective bargaining which is of relevance here is integrative bargaining which is concerned to solve problems of concern to both labour and management and thus produce mutual benefits, i.e. in terms of game theory there is a varying sum, as opposed to a fixed sum, pay off matrix. This particular bargaining sub-process was conceived to be a three phase affair: (i) problem identification; (ii) the generation of alternative solutions, and (iii) the selection from among these alternatives. The emphasis of the Walton and McKersie approach was very much on inter-personal and intergroup behaviour, with the problem solving orientation of integrative bargaining being seen to be facilitated by the following factors: a motivational orientation toward problem solving; the availability of necessary information, clarity of language, trust and a supportive overall union-management relationship. These necessary preconditions for joint problem solving would be aided by frequent contact between the parties so that problems could be considered as they arose, with an early discussion of items with problem solving potential.

The first attempt to test the validity of this problem solving subprocess of collective bargaining was that by Peterson and Tracy.(16) Their study sought to identify the conditions, behaviour and procedures that enhanced the problem-solving potential of negotiations by considering the following hypotheses:

(i) Success in problem solving is directly related to the cooperativeness of the pre-existing working relationship between the two negotiating teams.

(ii) Success in problem solving is directly related to recognition and approval from team members, constituents and opponents.

(iii) Success in problem solving is directly related to the extent to which negotiators regard themselves as professional negotiators.

(iv) Success in problem solving is directly related to the length of the bargaining relationship between the chief negotiators.

(v) Success in problem solving is directly related to the frequency of contact, as well as the openness of communication and information.

(vi) Success in problem solving is directly related to the effectiveness of one's own team's policy and administration.

These hypotheses were tested using questionnaire data obtained from the chief negotiator and two other negotiators on both the union and management sides involved in a sample of private sector negotiations during 1973. The results obtained confirmed the importance of all hypotheses, except (iii) and (iv) above. Their results also revealed the importance of a number of other hypotheses that were not derived from the original Walton and McKersie formulation- these additional hypotheses being largely concerned with the issue of relative bargaining power. In a related, but more broad ranging paper by the same authors it was further confirmed that perceived success in mutual problem solving was significantly influenced by the following factors: (i) the cooperative orientation of the other team; (ii) clarity and specificity of the other side; (iii) bargaining power; (iv) freedom of the negotiating team from constituent pressure; (v) friendliness of the other team. In short, the relevant influences were a mixture of economic, behavioural and procedural variables. In terms of Walton and McKersie's original formulation they noted that attitudional structuring (i.e. the obtaining and maintaining of a desired working relationship with the other party) was more closely tied to integrative bargaining than had been originally suggested. And, secondly, there was less direct conflict between the tactics used in integrative bargaining and those in distributive bargaining than had been originally implied.

The Peterson and Tracy study looked at problem solving within the process of contract negotiation, and not at the workings of problem solving structures and arrangements separate from the on-going collective bargaining process. Moreover, they used a rather 'soft' dependent variable (i.e. 'feelings of success' in joint problem solving) in their analysis. In a subsequent study of the operation of general purpose union-management problem solving committees Driscoll (17) used a more tightly defined dependent variable - i.e. the effectiveness of the committee being rated from (i) the perspective of one's own organisation (union or management), (ii) against the joint goals of mutual concern, and (iii) in relation to improvements in the collective bargaining process of negotiation and administration of agreements. The specific focus of the Driscoll study was to see if certain external conditions predicted problem solving behaviour and success over and above that predicted by the individual pre-dispositions that were so much emphasised in the original Walton and McKersie formulation. Accordingly, Driscoll hypothesised that the following sets of pre-conditions were likely to facilitate problem solving behaviour:

(1) <u>The External Environment</u>
- (a) An approximate equality of bargaining power between the two parties.
- (b) Outside pressure on the two parties to solve a common problem.
- (c) Harmonious relations between labour and management in their on-going relationships.
- (d) Support for problem solving behaviour by top leadership of both organisations.

(2) <u>Behavioural Predispositions of the Participants</u>
- (a) Positive interpersonal attitudes of trust, respect and liking between the negotiators.
- (b) Perceived importance of the committee.

(3) <u>Group Structure of the Committee</u>
- (a) Separation of problem solving efforts from on-going negotiations in terms of the people involved, the timing of discussions and the agenda.

These hypotheses were tested on the basis of questionnaire returns obtained from some 134 members of 38 different union-management committees. The results obtained indicated that positive interpersonal attitudes of trust, respect and liking, as originally hypothesised by Walton and McKersie, were strong predictors of problem solving, but over and above these factors external influences were also important. Specifically, if participants feel pressure from outside their organisation, if the environment makes for an approximate equality of bargaining power, if there is a history of harmonious relationships, and if the top leadership in both organisations supports the efforts, then labour-management committees are more likely to engage in joint problem solving behaviour and activities.

The only study which has developed the Walton and McKersie framework specifically to examine the effectiveness of joint health and safety committees is that by Kochan, Dyer and Lipsky.(18) Their study, which utilised questionnaire data obtained from the union and management representatives of some 25 such committees in upper New York State, examined three specific dependent variables: the frequency or continuity of interactions between unions and employers in established health and safety committees (i.e. frequency of committee meetings); the nature of the processes and strategies used to shape these interactions (i.e. problem solving or negotiating); and the determinants of the outcomes of these committees (i.e. number of recommendations from committees in the last six months). These three dependent variables can be interpreted as constituting necessary, if not sufficient, conditions for defining and measuring the effectiveness of such committees. As such they can be seen as something of a mid-point on a spectrum of measures for committee effectiveness, ranging from the individual perceptions of effectiveness held by committee members (the 'soft' end) through to perhaps the lowering of industrial accident rates at the 'hard' end. The independent variables used to try and explain variations in these three different measures of, or proxies for, committee effectiveness were grouped under the following sub-headings: (i) the degree of external pressure on the parties to deal with safety and health problems; (ii) the characteristics of the collective bargaining relationship; (iii) the organisational characteristics of the union; and (iv) the organisational characteristics of management. The basic findings of the study were as follows.(19)

It appears that if management perceives OSHA to be an important influence, deals with a strong union in a favourable labour relations climate, and is solidly committed to improving safety conditions in the plant, it will deal with the union in a problem-solving fashion. A co-operative approach by management, in turn, leads to a high level of committee activity, as measured by frequency of committee meetings (union perception) and number of recommendations which then is associated with the introduction of major safety improvements in the plant. The sequence for the union, however, is much more ambiguous and uncertain. Such factors as OSHA pressure and rank and file involvement will cause a high level of union activity with regard to safety and health issues. Union activity is characterised by the use of both problem-solving and negotiating strategies. But the activation of such a 'mixed' strategy does not necessarily result in the introduction of major safety improvements in the plant.

Major safety improvements appear to be less a function of union participation in a safety committee than of the direct pressure of OSHA regulations.

In the light of these findings the authors concluded that while the existence of external pressures on the parties does seem to provide the stimulus for the formation of a joint committee, the organisation policies of the union and the employers are at least as critical in determining the ultimate effectiveness of these joint efforts. Developing a strong commitment by top management and policies to implement that commitment, activating the interest and involvement of the rank and file union members, assigning representatives to the joint effort who have enough expertise and time to contribute to it, buffering the joint effort from the polemics of the collective bargaining process, and ultimately developing problem solving modes of interaction are critical steps that have to be achieved if joint change programmes are to be successful over any extended period of time.

It is especially necessary in the case of union-management committees to emphasise the importance of effective operation through time as such committees have typically been highly unstable institutions, with many tending to fall away through time. In Britain, for example, Marsh and Coker found a reduction of one-third in the number of joint production committees in federated engineering establishments between 1955 and 1961.(20) This sort of failure to stand the test of time means that any study of the effectiveness of joint health and safety committees conducted on a cross section or point of time basis should ideally have a follow up stage in order to see whether this effectiveness (and the factors responsible for it) have held up with the passage of time. The maintenance of commitment through time is explicitly considered in Kochan and Dyers'(21) model of organisational change in the context of union-management relations. In this regard they argue that for joint change initiatives to be successful and to survive over time the following obstacles will have to be overcome:

(i) Both parties must be able to see the change programmes as being instrumental to the attainment of goals valued by their respective organisations and/or constituents

(ii) The internal political risk to union leaders and management officials must be overcome.

(iii) The programmes must produce tangible, positive results in the short run and must demonstrate a high probability of being able to continue to achieve valued goals in the future.

(iv) The initial stimulus or pressure to embark on the programme must continue to be important and the initial goals of the programmes must continue to be of high priority to the parties.

(v) The gains or benefits from the initial programme must be equitably distributed among the workers and the employers.

(vi) The union must be perceived as being an instrumental force in achieving the goals or benefits of the programme and the union leaders must be protected from becoming overidentified as part of management.

(vii) The change process must be successfully integrated with the formal structure and procedures of contract negotiations and administration.

The potential role conflict for union leaders involved in joint problem solving initiatives is likely to be a particularly important determinant of the continuity of such arrangements, a fact well demonstrated in a recent study by Driscoll.(22)

Conclusions

In order to evaluate the effectiveness of joint health and safety committees one would ideally like to build up a detailed, systematic time series data set that would permit a statistical examination of the impact of committee activities on the industrial accident rate. This approach would be analogous to some of the 'impact of inspection' studies in the United States that will be discussed in the next chapter. However, as such data is unlikely to be available to researchers in the foreseeable future, a 'second best' approach would be to try and identify the characteristics of well functioning committees (and the factors responsible for this) that create at least a strong presumption of a favourable impact on accidents. A study along these lines should investigate both union and management perceptions of committee effectiveness (and the factors responsible for this) in order to see how closely they correlate. Furthermore, the study should attempt to see if outside researchers, who are not participant-observers of committee meetings, can develop objective measures of performance that create the presumption of effective operation, and then see how closely these indicators correlate with union and management perceptions of success. The analytical framework for such a study should be informed by the existing studies reviewed in the previous section. And finally, any such study should not be a one-off exercise, as a second or follow-up stage is essential to ensure that any revealed committee effectiveness is not of the 'hothouse' variety that is likely to fall away through the course of time.

NOTES

1. Report of the Safety and Health at Work Committee (Robens), 1970-72, Cmnd 5034, HMSO, London 1972, p.20-1.
2. ASTMS, Guide to Health and Safety at Work, p.41.
3. See, for example, Michael Cunningham, Safety Representatives: Shop Floor Organisation for Health and Safety: Studies for Trade Unionists, WEA, London, March 1978.
4. D A Castle, "Safety Representatives and Safety Committees - Power without Responsibility", New Law Journal, September 28, 1978, p.944-6.
5. Labour Research Department, LRD Guide for Safety Representatives, London, June 1978, p.31.
6. Michael Armstrong, Case Studies in Personnel Management, Kogan Page, London, 1979, p.202-6.
7. P B Beaumont and D R Deaton, "The Enterprise Response to Industrial Relations Legislation. The Case of Joint Health and Safety Committees", Industrial Relations Journal, March-April 1981.
8. See, for example, Karen Legge, Power, Innovation and Problem Solving in Personnel Management, McGraw Hill, London, 1978, p.69-71.
9. See, for example, John Gennard and M D Steuer, "The Industrial Relations of Foreign Owned Subsidiaries in the United Kingdom", British Journal of Industrial Relations, vol.IX, no.2, July, 1971, p.153-5.
10. Industrial Relations Review and Report, No.147, March 1977.
11. Industrial Relations Review and Report, No.189, May 1979.
12. Ministry of Labour, Works Safety Committees in Practice - Some Case Studies, HMSO, London, 1968.
13. Cited in J L Williams, Accidents and Ill-

Health at Work, Staple Press, London 1960, p.200.

14. Williams, op.cit., p.205.

15. Richard E Walton and Robert B McKersie, A Behavioural Theory of Labor Negotiations, McGraw Hill, New York, 1965, p.126-60.

16. R B Peterson and L N Tracy, "A Behavioural Model of Problem Solving in Labour Negotiations", British Journal of Industrial Relations, vol. XIV, No.2, July 1976. See also Richard B Peterson and Lane Tracy, "Testing a Behavioural Theory Model of Labor Negotiations", Industrial Relations, vol.16, No.1, February 1977.

17. James W Driscoll, "Problem Solving Between Adversaries: Predicting Behaviour in Labor-Management Committees", Mimeographed Manuscript, Sloan School of Management, Massachusetts Institute of Technology.

18. Thomas A Kochan, Lee Dyer and David P Lipsky, The Effectiveness of Union-Management Safety Committees, The W E Upjohn Institute for Employment Research, Kalamazoo, Michigan, 1977.

19. Kochan, Dyer and Lipsky, ibid, p.72.

20. A I March and E E Coker, "Shop Steward Organisation in the Engineering Industry", British Journal of Industrial Relations, July 1963.

21. Thomas A Kochan and Lee Dyer, "A Model of Organisational Change in the Context of Union-Management Relations", Journal of Applied Behavioural Science, Vol.12, No.1, February 1976.

22. Jim Driscoll, "Coping with Conflict: An Exploratory Field Study of Union-Management Cooperation", International Journal of Applied Psychology, 1980.

Chapter Seven

UNION INVOLVEMENT AND WORKPLACE HEALTH AND SAFETY IN THE UNITED STATES

This chapter will outline in rather general terms a number of issues and arguments that have surrounded the operation of the Occupational Safety and Health Act of 1970 (henceforth referred to as OSHA) in the United States, on the grounds that such matters are likely, at least to a certain degree, to be increasingly heard in Britain in the years to come. More specifically, we consider the major implications of OSHA for union-management relationships with a view to drawing out a number of behavioural and research insights likely to be of potential relevance in Britain. In this regard the operation of grievance procedures in handling workplace health and safety problems and union activities in relation to occupational diseases, as opposed to industrial accidents, are stressed as being particularly important. These matters are among those taken up in the concluding chapter where we consider *the* issues for the future in this subject area in Britain.

The reason for the sole concentration on the United States in this chapter is the strong historical and substantive parallel between Britain and the United States with regard to the extent and nature of union involvement in the area of workplace health and safety. As in Britain, the basic responsibility for plant health and safety matters has long been in the hands of management in the United States, with the unions generally being, as Chamberlain and Kuhn put it, "... content to leave responsibility with management, reserving the right to prod and criticise where they believe provision has been inadequate."(1) However, OSHA provides important new rights of information and discussion to employees and their representatives and as such offers considerable opportunity for substantial union involvement in the preventative aspects of workplace health and safety. Accordingly, a review of the operation of these employee and union rights should help us identify some of the issues and problems likely to be of interest and relevance in Britain where a not too dissimilar set of rights have more recently become law. The value of a comparison with the United States is also enhanced by the fact that some academic analysis of the operation and impact of arrangements for union involvement in workplace health and safety has already been undertaken there, although a great deal more analysis certainly remains to be undertaken.(2) In contrast, for other countries, with admittedly more extensive rights of employee and union involvement in workplace health and safety, there exists little more than descriptive outlines of the relevant institutional arrangements.(3)

The Pre-OSHA Position

The union movement in the United States included amongst its earliest demands the (i) guarding of hazardous machinery and correction of other serious workplace hazards, and (ii) compensation for industrial accident victims and their dependents, particularly in cases of serious, permanent disablement or death. (4) The State of Massachusetts enacted the first law requiring the safeguarding of machinery, hoists and elevators in 1877, while New York enacted a factory safety law in 1887 and added safety inspection to the duties of factory inspectors ten years later. At the time of the seventh annual convention of the International Association of Factory Inspectors in Chicago in 1893 there were 14 states and provinces with factory laws and 110 inspectors.(5) In 1911 the state of Wisconsin adopted the famous 'blanket' safety provision which superseded all existing safety laws, and established an Industrial Commission to produce separate safety codes in the form of administrative orders.(6) This approach spread rapidly to other states, although the resulting provisions were held to vary quite widely in terms of their effectiveness. There were continuing initiatives in the area of workplace health and safety regulations by individual state governments right through until the end of the 1960s. The resultant body of state government based legislation was characterised by the adoption of safety codes, frequently based on the standards set by various voluntary safety organisations, and an emphasis on investigation and consultation rather than the use of sanctions.(7) The Federal Government had assumed, at various stages, whole or partial responsibility for the safety of interstate railroad workers, air transport employees, seamen and coalminers, but more generally its role was limited to the provision of information and technical assistance through the medium of the Department of Labor.

In regard to accident compensation most state governments enacted some form of employers liability law between 1895 and 1910 and in 1908 the Federal Employees Liability Act was passed to provide protection for employees of common carriers engaged in interstate commerce.(8) Although such laws were considered an improvement over the earlier common law protection for employees they were still considered to be deficient, to the extent that by 1916 31 states and the federal government had appointed fact finding bodies to study their effectiveness. The outcome of such investigations was the passage of workmens compensation laws by some 22 states in the period 1910-14, with the remainder following suit in subsequent years; the last state to do so being Mississippi in 1948. There were considerable differences in the technical characteristics and extent of protection provided by these state laws; in nearly half of the states, for example, the compensation laws were elective in the sense that employers could refuse to be covered by them, although in such cases they typically lost their right of common law defence in the event of a damages suit. Indeed in 1972 the National Commission on State workmens compensation laws reported that only 22 of the 50 states met even half of the 'effectiveness' criteria for such laws which were suggested by the Department of Labor.(9) The Commission's report urged the need to extend coverage to all workers, to increase benefit levels, remove money and time limits on compensation and provide a fuller coverage of occupational diseases.

In the United States unions have, according to Kochan, Dyer and Lipsky,(10) historically followed a three-pronged strategy for dealing with the safety and health needs of their members: (1) lobbying for the passage and enforcement of state and federal legislation; (2) negotiating collective bargaining provisions to protect and/or compensate workers for risks associated with hazardous jobs; and (3) establishing joint union-management plant committees to monitor and improve safety and health conditions on a continuous basis. In relation to the first medium of influence Barbash has suggested that "the unions have been, perhaps, the most active element in enlisting the aid of federal and state government to provide safe and healthful workplaces."(11) The major illustration of this activity cited by Barbash was the United Mine Workers pursuit of an enforceable federal inspection law in the mines, a goal finally achieved in 1952. In considering the second medium of union influence the bargaining structure hypothesis which was outlined in Chapter 3 would lead us to expect relatively more collective bargaining over health and safety matters in the relatively decentralised bargaining system (i.e. company and plant based) of the United States than was evident in the traditional, industry based structure of Britain. The effects of this structural influence should have been substantially enhanced by the ruling of the National Labor Relations Board in 1966 that industrial safety was a mandatory subject for collective bargaining.(12) There were certainly health and safety provisions in collective agreements prior to the passage of OSHA, the extent and nature of such provisions being illustrated by the contents of Table 7.1.

The contents of Table 7.1 reveal that relatively few of these health and safety provisions appeared in even half of the collective agreements examined. Furthermore, the most numerous provisions were very much those that consisted of little more than a general statement of policy or a pledge of compliance with existing laws, whereas the more substantive provisions - i.e. joint safety committees, safety inspections, discipline for non-compliance, employee and union rights with regard to safety - were represented in less than one-quarter of the total number of collective agreements examined. The impression, which is conveyed by the figures in Table 7.1, that health and safety matters were far from prominent subjects of joint decision making in the pre-OSHA years, is supported by a number of case studies. The following findings for the chemical industry are perhaps not untypical in this regard.(13)

> There is no evidence that collective bargaining had any discernible impact on the safety and health policies and programs of employers in the chemical industry prior to the passage of the Act. None of the firms studied reported having faced serious union demands in the safety and health area in contract negotiations prior to 1970, and few reported strong union

Table 7.1: General Provisions Referring to Safety and Health in Major Collective Agreements Before 1971

Provision	Agreements	Workers Covered
All agreements studies	503	1,651.7
Total referring to safety and health	476	1,582.4
General policy statements	200	551.2
Union-management co-operation pledges	98	286.4
Safety committees	109	329.8
Joint safety committees	93	273.1
Safety inspections	63	270.2
Employer pledges of compliance with the law	204	751.6
Employee compliance with safety rules or laws	197	660.5
Discipline for non-compliance	88	298.1
Employee rights with regard to safety	112	438.5
Union rights with regard to safety	75	292.5
Safety equipment	262	831.6
Sanitation provisions	247	847.5
Physical examinations	104	309.9
Accident procedure or compensation	300	925.4

Source: Winston Tillery, "Safety and Health Provisions Before and After OSHA", Monthly Labor Review, September 1975, p.42.

pressure for the creation of formal union-management safety committees prior to the passage of the Act. There were isolated cases of demands for wage-rate adjustments based on alleged hazardous working conditions and of grievances over unsafe conditions, but no widespread pressure for hazard premiums or hazard elimination. Most of the cases of demands for hazard premiums were seen by management as involving situations of unpleasant, rather than unsafe or unhealthful working conditions. Similarly, some of the grievances involving safety issues arose in conjunction with changes in work procedures or crew sizes and reinforced complaints about effort and staffing rather than about safety.

There were certainly exceptions to the sort of picture described above, most notably in the high risk, steel industry, but in general the available evidence points to the fact that health and safety matters did not figure very prominently as a subject of joint discussion, much less one of joint decision making, in the period prior to the passage of OSHA. Joint health and safety committees are among the provisions listed in Table 7.1, but as Kochan, Dyer and Lipsky specifically distinguished them as a separate route of union influence it is worth considering the extent of their development in a little more detail.

The American industrial safety movement is typically held to have originated in 1907 with "the stimulation of safety work through the vigorous campaigns launched by the United States Steel Company, the Chicago and Northwestern Railroad, the International Harvester Company, and a few other outstanding corporations"(14) being considered of substantial importance. This movement was from its industrial engineering origins highly paternalistic in nature with relatively little priority being accorded to the obtaining of systematic employee and union involvement in the subject area.(15) And this orientation was very much reinforced by the legal system which made workplace health and safety very much a management responsibility; the result was that workplace health and safety matters for long figured prominently in lists of managements' 'reserved rights'.(16) As in Britain, attempts to encourage the establishment of joint health and safety committees can be seen from the early years of this century. The Chicago and Northwestern Railroad was particularly important in demonstrating the potential value of such committees,(17) but the available evidence clearly points to the relatively limited and uneven development of this particular form of health and safety arrangement. One such body of evidence, for the period of time immediately prior to the introduction of OSHA, is set out in Table 7.2.

The aggregate figures in Table 7.2 indicate that the establishment and workforce coverage of safety committees was far from extensive, and with considerable inter-industry variation being apparent.

Table 7.2: Labor-Management Health and Safety Committees in Agreements Covering 5,000 Workers or More in the United States, by Industry, 1970

Industry	Agreements providing for a Safety Committee (%)	Workers covered by a Safety Committee (%)
Ordnance and accessories	50.0	35.5
Food and kindred products	16.7	10.2
Printing, publishing and allied	25.0	35.3
Rubber and miscellaneous	75.0	87.3
Stone, clay and glass	75.0	85.2
Primary metals	100.0	100.0
Machinery, except electrical	33.3	37.1
Electrical machinery	20.0	15.4
Transport equipment	46.2	17.8
Transportation	59.1	69.3
Utilities: electricity and gas	66.6	79.3
Construction	6.7	3.9
Manufacturing	39.7	34.0
Non-manufacturing	16.7	24.3
All industries	28.2	29.9

Source: US Department of Labor, Characteristics of Agreements Covering 5,000 Workers or More, Bulletin 1686, 1970, p.17.

These sorts of figures do help place our Chapter 3 discussion of the limited and uneven development of joint health and safety committees on a voluntary basis in Britain in some sort of comparative context. It is to be expected that the extent of establishment of joint health and safety committees will have increased considerably since the passage of OSHA. This is a matter that we examine later in this chapter. However, at least one union official has argued that "... combining the OSHA complaint procedure with the existing grievance procedure is more effective than the joint safety committee. This approach avoids many of the delays associated with the joint safety committees."(18) This is also a matter taken up in subsequent sections of this chapter.

The Passage of OSHA

As early as the 1950s there were indications of a growing demand among some union leaders for an increased Federal Government role in the area of workplace health and safety.(19) However, these demands increased significantly in the late 1960s due to the following factors:(20)

(i) The reversal of the downward trend in the industrial accident rate in the manufacturing sector.(21)

(ii) The increased realisation that occupational health was being endangered by little understood air contaminants and other physical or chemical agents in the work environment, including the highly publicised 'black lung' disease in coalmining.

(iii) The increased attention that the consumer movement led by Ralph Nader began to give to public and occupational safety and health matters.

(iv) The increased evidence of worker concern with health and safety and health matters (22) that forced the union leadership to rethink their priorities.

(v) The November 1968 Mannington coalmine disaster which aroused miners, consumer advocates, and political activists to a new pitch of concern about safety and occupational health, coming as it did at a time of increasing concern about coal-miners pneumoconiosis (black lung).

These demands were eventually met by the passage of OSHA in 1970. This Act should be seen in the larger context of a shift away from the traditional policy concern of regulating wages, hours and working conditions indirectly through regulating the process of collective bargaining towards a more direct regulation of specific terms and conditions of employment. This policy shift, which is well evidenced by the increase in the number of regulatory programmes administered by the Department of Labor from 40 in 1960 to 134 in 1975,(23) has come in for considerable general criticism, with OSHA being to the forefront in this regard. However, before we examine some of the criticisms that have been levelled at OSHA it is necessary to provide a brief outline of its major terms and provisions.(24)

The provisions of OSHA cover some 57 million employees in more than 4.6 million establishments throughout the country that are engaged in interstate commerce. It is only public sector employees and private sector employees, such as miners, who are covered by other safety and health acts, that are excluded from the OSHA provisions. The Occupational Safety and Health Administration, which is located in the Department of Labor, with an annual budget of some $70 million is responsible for setting and enforcing standards. The National Institute for Occupational Safety and Health (NIOSH), which is located in the Health, Education and Welfare Department, was established at the same time as OSHA. This body, which has a budget of nearly $30 million is responsible for conducting research and recommending criteria on health and safety standards to OSHA, undertaking education and training programmes, and performing work in the area of safety engineering. A three-member Occupational Safety and Health Review Commission has been established to handle the appeals of employers on violations and penalties. Finally the Act provides that the individual states may assume responsibility for employee safety and health if they submit plans that are 'at least as effective' as the federal plan. A number of states have already submitted such plans and had them approved.

The Benefits and Costs of the Operation of OSHA

OSHA seeks to achieve a reduction in industrial accident rates through the issuance of legally enforceable minimum safety (and health) standards.

This basic standards-enforcement strategy has been argued to have a number of *inherent* weaknesses as a method for preventing injuries at the workplace.(25) In this regard it has been claimed that relatively few workplace injuries are caused by violations of standards, and even fewer by violations that inspectors can detect. Furthermore, the particular standards issued under OSHA have in the main been existing 'consensus' standards, derived in the main from voluntary safety organisations, which were designed as voluntary guidelines or benchmarks rather than as legislative minima which must be complied with. These 'inappropriate' standards are alleged to have encouraged high rates of employer non-compliance; an effect further encouraged by the relatively low costs of non-compliance. The issue of the costs of non-compliance was central to the work of Viscusi who developed a model which considered the financial incentives created by OSHA for firms to improve health and safety.(26) The basic data from this exercise is set out in Table 7.3.

The expected penalty amount, which can be imposed by OSHA, is the product of the probability of an inspection, the number of violations per inspection, and the average penalty per violation, but as the contents of Table 7.3 reveal none of these is of very large magnitude. The average number of inspections per enterprise, for instance, was only 0.079 over the five year period or approximately 0.02 on an annual basis, while the average number of standards violations cited per inspection was only 3.7, and the average penalty per violation only $25.68 from 1971-75. These sorts of figures suggested that OSHA's present penalty levels were too low to create an effective financial incentive to improve workplace health and safety, and perhaps not surprisingly Viscusi found no evidence of any significant OSHA impact on industry health and safety investments and injury rates for the period 1972-75. The only studies which have identified a positive, OSHA effect have used *disaggregated* injury rate data. A recent paper by Smith,(27) for example, argued that OSHA may reduce accidents due to effects on uninspected firms (e.g. through threat effects) as well as inspected firms. However, any positive effect is more likely to occur in the latter case but inspections cover the workplaces of at most 10 per cent of employees each year so that any reduction in accident rates induced by OSHA inspection activities will be too small to be identified on the basis of aggregate injury rate data. In a similar

Table 7.3: OSHA Activities and Expenditures, 1971-75

	1971	1972	1973	1974	1975
Total budget (in thousands)	$25,537	$52,629	$69,891	$86,207	$109,594
Enforcement budget (in thousands)	$10,172	$19,235	$38,543	$66,991	$ 88,287
Inspections	14,500	36,100	67,153	79,605	88,801
Compliance Rate	28%	28%	26%	24%	21%
Violations	35,800	125,400	224,786	308,702	367,401
Citations	9,500	23,900	43,099	58,338	68,955
Proposed Penalties (in thousands)	$ 738	$ 3,121	$ 6,059	$ 6,950	$ 10,411

Source: W Kip Viscusi, "The impact of occupational safety and health regulation", The Bell Journal of Economics, Vol.10, no.1, Spring 1979, p.125.

vein, Mendeloff(28) found no OSHA impact on the overall US or California injury rates, but when the California injury rate data was disaggregated by accident type it was found that the rate for the category that OSHA enforcement was most likely to affect was in fact significantly lower than predicted. These findings led Mendeloff to conclude that: (29)

> The most important insight to be derived from an examination of OSHA's effect on injuries does not pertain to the estimate of its quantitative impact, but to its potential scope. Several studies indicate that only a small minority of all injuries are caused by violations that inspectors could detect on scheduled inspections. OSHA may actually be achieving a substantial part of its potential; the problem is that the potential of its present program is so small. Although small improvements are possible within the current emphasis on seeking detectable violations, the more important gains to be made require expanding OSHA's scope and preventing injuries more cost effectively.

The mention of a more cost effective approach to accident prevention requires further consideration as many of the critics of OSHA have charged that it has brought little in the way of significant gains in worker health and safety, but that its operation has involved substantial private and social costs. A recent study (30) of Federal Government expenditure on regulatory activities reported that expenditure on job safety (and other working conditions) had risen as follows: from $62 million in 1970 to $310 million in 1974, to $496 million in 1978 and an estimated $678 million and $701 million in 1979 and 1980 respectively. This same study reported that in the calendar year 1976 the administrative cost of regulating job safety (and other working conditions) was $483 million, but the compliance cost was estimated to be something of the order of $4,015 million. The particular criticism levelled against OSHA is that its procedures have been formulated with minimal concern for resource costs, i.e. the agency has only considered compliance with its standards in terms of technical feasibility, and not in terms of cost effectiveness. OSHA's unwillingness to consider the potential value of personal protective devices as a means of meeting exposure limits has been especially singled out as an example of their relative lack of concern with the criterion of cost effectiveness.(31) On the basis of their review of the benefits and costs of OSHA's operation Nichols and Zeckhauser questioned whether,(32)

> ... the agency's failures have resulted simply from faulty execution (including the overly hasty adoption of thousands of consensus standards, excessive emphasis on safety relative to health, the inevitable start-up problems of any new agency, and, more controversially, the exclusion of economic considerations in all but extreme cases) or whether they were inherent in the basic approach taken: direct regulation through standards and inspections.

In their view there are substantial _inherent_ limitations to the standards setting-inspection approach adopted by OSHA which means that it needs to be complemented, or better still replaced, by an approach embodying (i) the expanded provision of information and (ii) a greater use of financial incentives. Under the latter heading, for instance, they propose an injury tax on employers which they argue would provide firms with a more powerful incentive to improve their safety performance by considering the whole range of factors that contribute to accidents, and not simply the limited number of physical conditions susceptible to direct regulation. This tax approach would appear to be less suitable, at least on administrative grounds, for dealing with occupational health problems due to the long latency periods of occupational diseases which make it difficult to connect individual cases of disease and illness with particular firms. In the case of occupational health problems Nichols and Zeckhauser therefore put forward a modified tax proposal, namely a tax on workers exposure to health risks, that is analogous to the use of effluent charges for environmental pollutants. However, such tax proposals are typically held to be politically unattractive and thus unlikely to be introduced in the immediate future. For this reason Mendeloff, (33) for example, sets out a number of 'more realistic' policy reforms for OSHA that are designed to increase the proportion of injuries that enforcement can affect, and improve the cost effectiveness of any resulting reduction in accidents. On the other hand, Lawrence Bacon has argued that collective bargaining has the greatest potential for remedying the structural deficiencies of OSHA's operation.(34) It is the subject area of this latter proposal that we now turn to consider.

The OSHA Impact on Union-Management Relations

Under OSHA the five important rights provided for the employee or employees representative are as follows:-

i) The employee who believes that a violation of a job safety or health standard exists which threatens physical harm or that an imminent danger exists, may request an inspection by sending a signed written notice to the Department of Labor (his or her name is not revealed to the employer when the employee requests anonymity). Under the provisions of the Act, the employee may not be discharged or discriminated against for filing such a complaint.

ii) When the Department of Labor inspector arrives an employee representative may accompany the compliance officer on his or her visit for the purpose of aiding the inspection. This is the so-called 'walk-around right'.

iii) If the employer is cited by OSHA and protests either the fine or the abatement period, the employee is entitled to participate in the hearing and to object to the period of time fixed in the citation for abatement of the violation.

iv) With respect to exposure to toxic materials or other physical harmful agents, the employee may observe the company's monitoring process. If the substance is defined as toxic and covered by an OSHA standard, the employee is entitled to information about his or her exposure record.

v) The employees authorised representative (rather than the employee) may request that the Secretary of HEW determine "whether any substance normally found in the place of employment has potentially toxic effects in such concentrations as used or found." If HEW makes this finding, the Secretary of Labor may institute a rule making procedure to set a safe exposure level for that substance.

These rights have the potential for considerably extending union influence in the preventative aspects of workplace health and safety. In this regard Ashford has suggested that unions are seeking to build specific clauses into collective agreements which cover the following matters:-(35)

i) Funding: a joint labor-management trust fund specifying a per employee-hour contribution by the employer. This fund would provide a financial base for conducting industry wide studies into the causes and prevention of occupational injuries and diseases.

ii) Expanded protection: incorporation of OSHA regulations and standards into labor contracts, as well as health and safety provisions based upon information presented by the union.

iii) Arbitration: provisions for an expedited arbitration procedure when occupational health and safety disputes occur. Clauses guaranteeing the right to strike over alleged dangerous conditions are also being sought by the unions. These provisions place full responsibility for resolving occupational health and safety problems directly on the parties most concerned.

iv) Union inspection rights: provisions for unlimited union inspection rights.

v) Use of impartial experts: provisions for the use and funding of impartial health and safety experts when disputes arise or when conditions become dangerous.

vi) Training of union health and safety stewards: provisions for on the job health and safety training for selected union members to aid in the enforcement of standards.

vii) Multiphasic screening tests for employees: periodic medical examinations to help identify health hazards to which union members are exposed.

viii) Establishment or expansion of safety and health committees: to conduct inspections and investigations, and to participate in the setting of general policy.

Earlier, in Table 7.1 we provided some evidence on the extent of health and safety provisions in collective agreements prior to the passage of OSHA. As a comparison with those figures we present below in Table 7.4 the extent of such provisions in collective agreements almost immediately following the passage of OSHA.

The figures in Table 7.4 when compared with those in Table 7.1 reveal that all of the health and safety provisions listed became more numerous after OSHA went to effect, although the extent of the increases in this early period of time were generally slight or moderate. Predictably, more sizeable changes have come about in recent years. For example, only 31 per cent of the collective agreement contracts in the Bureau of National Affairs sample in 1970 provided for health and safety committees, compared to 43 per cent in 1978, with comparable sized increases occurring in the frequency of other health and safety provisions over the same period of time,(36) There also appears to have been something of an increase in safety related strikes in the post OSHA years; in 1969 such strikes accounted for only 1.4 per cent of all stoppages whereas in the years 1973, 1974 and 1975 the relevant figures were 2.9, 2.5 and 3.3 per cent.(37)

This evidence of increased joint involvement in workplace health and safety matters has not been randomly distributed across the whole range of union-management relationships. The especially active unions in this area are the United Steel Workers of America, the United Automobile Workers, the United Paperworkers International Union, the International Association of Machinists, the United Rubber Workers, the Boilermakers Union, the United Mine Workers, and the Oil, Chemical and Atomic Workers International Union.(38) For example, the Oil, Chemical and Atomic Workers Union successfully negotiated the following provisions in collective agreements in 1973: (i) joint labor-management health and safety committees with power over workplace conditions; (ii) periodic health surveys of plants by independent health consultants approved by both labor and management; (iii) medical examinations of workers at company expense; and (iv) the provision of annual employee morbidity and mortality data by the company. In the same year the United Auto Workers successfully negotiated a six-point health and safety programme in the automobile industry. The UAW contract included: (i) the establishment of a new hierarchy of full-time company-

Table 7.4: General Provisions referring to Safety and Health in Major Collective Agreements, after 1971

Provisions	Agreements	Workers covered
All agreements studied	503	1,811.1
Total referring to safety and health	487	1,767.3
General policy statements	219	607.3
Union-management co-operation pledges	107	317.2
Safety committees	124	434.5
Joint safety committees	108	362.3
Safety inspections	77	342.8
Employer pledges of compliance with law	259	1,002.0
Employee compliance with safety rules or laws	246	871.0
Discipline for noncompliance	124	491.6
Employee rights with regard to safety	134	542.1
Union rights with regard to safety	82	383.6
Safety equipment	291	1,116.9
Sanitation provisions	252	1,034.7
Physical examinations	107	401.8
Accident procedures or compensation	324	1,085.9

Source: Winston Tillery, "Safety and Health Provisions Before and After OSHA", *Monthly Labor Review*, September 1975, p.42.

paid UAW safety and health representatives; (ii) continuation of the prior union right to strike over safety and health issues; (iii) provisions for regular weekly or bi-weekly inspections by UAW representatives or joint labor-management plant committees; (iv) UAW rights to sampling and monitoring equipment for noise and gas, to be provided by the companies; (v) supporting equipment for other environmental surveillance conducted by the union; (vi) union access to detailed data from company and government records on employee exposure to dangerous physical agents and chemicals; (vii) union access to detailed data on harmful physical agents and chemicals; and (viii) company paid physicals and other examinations and tests for employees exposed to harmful physical and chemical agents. More generally, a study by Wendling (39), which was drawn on in the analysis of Chapter 3, sought to account for the relative number and importance of health and safety provisions in nearly 600 collective agreements in effect at mid-1974 in the manufacturing sector. The potential explanatory variables were grouped under three basic sub-headings, designed to test the hypotheses that health and safety provisions would be more numerous and important when (1) the industrys' workforce is exposed to numerous hazards, (2) the industrys' workforce is highly unionised, and (3) the composition of the internal workforce is relatively homogeneous. The major substantive result to emerge from this exercise was the positive relationship between the level of workplace hazards and the number and/or value of safety provisions included in collective bargained agreements.

The general provisions reported in Tables 7.1 and 7.4 which state that the employer is responsible for providing a safe work environment or that the employer agrees to comply with applicable safety laws allow the union to use the contractual grievance procedure to process safety complaints. The potential importance of this fact is suggested by the earlier noted comment from one union official that the grievance procedure is more effective than joint safety committees in dealing with health and safety complaints as it avoids the delays associated with the latter form of arrangement. However, some figures cited by Geisung suggest that health and safety grievances are still a very small proportion of grievances; for example, the grievance records of General Motors Corporation revealed that 0.7% of cases in 1974 and 1.5% in 1975 were safety or health related.(40) It was noted earlier that unions are seeking provisions for an expedited arbitration procedure when health and safety disputes occur. The experience in the basic steel industry is interesting here. According to union figures, some 186 health and safety related grievances have gone to arbitration since 1966. The union was successful in 73 of these disputes, with the success rate varying considerably according to the particular issue in dispute.(41) An increasingly important issue is the right of a group of employees to refuse to work _in protest_ against allegedly hazardous working conditions, as whether such action is a protected activity that is not subject to the usual no-strike contract clause is a debatable issue. The only major exception to the accepted rule of 'work now, grieve later' is provided by Section 502 of the Labor-Management Relations Act which states that: "nor shall the quitting of labor by an employee or employees in good faith because of abnormally dangerous conditions for work at the place of employment of such employee or employees be deemed a strike under this Act." A detailed study of this matter by Summa led him to conclude that "if employees want to avoid discipline for a work stoppage on the grounds that their health or safety is endangered, they must meet three criteria: immediate involvement; exhuastion of any applicable contractual provisions; and proof, not that their feelings were bona fide or even reasonable, but that there was an actual danger to their health or safety."(42) Finally, OSHA holds employers responsible for seeing that their employees wear prescribed safety equipment so that an increase in the disciplining of employees for noncompliance with safety work rules, with its potential for an increase in grievances, is likely to result.(43)

The section of the Act with the most obvious impact on industrial relations is Section 8(F)(1) which gives any employee (or union) the right to complain to the Secretary of Labor when the employee (or union) believes that a violation of a safety-health standard threatens physical harm or that an imminent danger exists. This right to initiate an inspection has been held by some to have considerable potential for 'misuse' in the sense that the provisions could be utilised for _tactical_ purposes to bring pressure on employers to concede other negotiating demands and grievances. For example, in support of working rule demands the unions could question the manpower requirements on new machinery on the grounds of their 'adverse' safety implications. A number of examples of the tactical use of OSHA were cited in a recent study of the Act's impact in the chemical industry.(44) More generally, however, the fact that just under 95% of all complaints in the years 1973-76 were judged to be valid strongly suggests that such 'misuse' of the provisions is relatively rare.(45)

The issues on the frontier of bargaining in this area are very much those of occupational health rather than industrial safety. This is the result of a slowly accumulating body of knowledge about the extent and effects of exposure to toxic substances at the workplace. The fact of the matter is that (46),

> About 21 million Americans - or one out of every five workers - are exposed to hazardous substances on the job, and more than twice that many are exposed to such substances some time during their working life. One out of four Americans will suffer from cancer during his or her lifetime, and up to 38% of all cancers in the United States are related to substances in the workplace. Between 8 and 11 million workers have been exposed to one cancer causing substance, asbestos, since the Second World War, with more than 2 million of those expected to die from asbestos-related cancer. Two million workers face exposure to benzene, a highly toxic chemical which can cause leukaemia. Another 1.5 million workers are exposed to arsenic, also a cancer threat. Three out of four coalminers now receiving a pension have an irreversible lung disease caused by coaldust. About 800,000 cases of job-related skin disorders occur each year as the result of exposure to toxic substances.

Some figures cited by Viscusi indicated that a number of collective agreements specify wage premiums for particular on the job hazards; for example, 109 contracts out of the total sample size of 1,724 had hazard premiums for working with acid, fumes or chemicals.(47) However, recent initiatives have moved beyond this compensatory approach to concern themselves with prevention or abatement measures. Some recent examples of such initiatives are as follows:(48)

i) The Oil, Chemical and Atomic Workers Union have

negotiated periodic inspections by independent experts approved by both the union and employer, company paid medical examinations, and the provision to workers ot annual information on illnesses.

ii) The United Auto workers have successfully demanded that the automobile manufacturers provide information on toxic substances in the workplace, and provide equipment for local union representatives to test for exposure to hazards.

iii) The United Rubber workers and the major rubber producers have established an employer- financed joint research programme in co-operation with two major universities.

iv) The United Steel workers have negotiated provisions in collective agreements on the maximum loss of pay that can be suffered by a worker who is transferred to a new job because of over-exposure to lead.

These examples are certainly not representative of current practice in the United States, but they are useful in providing some indication of the basic strategies that are likely to be increasingly looked to in the future. In Kochan's view there are essentially two emerging strategies- (i) compensation for the accumulated effects of past exposures by transferring workers to healthier jobs without any loss in pay; and (ii) reducing the exposures of workers in the future. Under this latter heading one would find support for tighter Government standards, improving information flows about toxic substances and negotiating provisions for more industrial hygienists, examinations, etc.(49)

Conclusions

In this chapter we have reviewed at some length a number of the major issues and questions that have surrounded the operation of OSHA in the United States. The potential value of the exercise follows from the interest of these issues in their own right together with the likelihood that at least some of them will be increasingly raised in relation to the operation of the Health and Safety at Work Act in Britain. Accordingly this review of American experience is combined with the analysis and results of earlier chapters to provide a list of a number of the important issues concerning union involvement in workplace health and safety matters which should be considered by researchers and policy-makers in Britain in the future. This list is the subject of our concluding chapter.

NOTES

1. Neil W Chamberlain and James W Kuhn, Collective Bargaining, McGraw-Hill, New York, Second Edition, 1965, P.104.

2. See Labor Management Relations Research Priorities for the 1980s, Final Report to the Secretary of Labor, US Department of Labor, January 1980, p.33-4.

3. See, for example, Paul Versen, "Employer-Worker Co-operation in the Prevention of Employment Injuries", International Labour Review, Vol.118, No.1, January-February 1979.

4. Ronald P Blake (ed.), Industrial Safety, Prentice Hall, Englewood Cliffs, 3rd Edition, 1963, p.13.

5. Cited in John R Commons and Associates, History of Labor in the United States, 1896-1932, Vol.III, Macmillan, New York, 1935, p.366.

6. Herman M Somers and Anne R Somers, "Industrial Safety and Health in the United States", Industrial and Labor Relations Review, Vol.6, No.4, July 1953, p.483.

7. Leo Teplow, "Comprehensive Safety and Health Measures in the Workplace", in Joseph P Goldberg, et. al. (eds.), Federal Policies and Worker Status Since the Thirties, IRRA, Wisconsin, 1976, p.211.

8. This paragraph is based on Sanford Cohen, Labor in the United States, Charles E Merrill, Ohio, Third Edition, 1970, p.551.

9. Cited in David M Kasper, "An Alternative to Workmen's Compensation", Industrial and Labor Relations Review, Vol.28, No.4, July 1975.

10. Thomas A Kochan, Lee Dyer and David B Lipsky, The Effectiveness of Union-Management Safety and Health Committees, The W E Upjohn Institute for Employment Research, Michigan, 1977, p.1.

11. Jack Barbash, The Practice of Unionism, Harper and Brothers, New York, 1956, p.250.

12. Herbert R Northrup, Richard L Rowan and Charles R Perry, The Impact of OSHA on Three Key Industries - Aerospace, Chemical and Textiles, Industrial Relations Research Unit, The Wharton School, University of Pennsylvania, 1978, p.194.

13. Northrup, Rowan and Perry, op.cit., p.195.

14. Commons and Associates, *op.cit*., p.367.

15. Somers and Somers, *op.cit*., p.476-80.

16. See, for example, Harold W Davey, *Contemporary Collective Bargaining*, Prentice Hall, Englewood Cliffs, 1972, p.107.

17. Commons and Associates, *ibid*., p.368.

18. John Zalvsky, "The Worker Views the Enforcement of Safety Laws", *Labor Law Journal*, Vol.26, No.4, April 1975, p.225.

19. Somers and Somers, *ibid*., p.485.

20. Teplow in Goldberg et.al.(eds.) *op.cit*., p.222.

21. James R Chelius, *Workplace Safety and Health: The Role of Workers Compensation*, American Enterprise Institute for Public Policy Research, Washington, 1977, p.12-15.

22. R P Quinn, et.al., "Evaluating Working Conditions in America", *Monthly Labor Review*, Vol.96, 1973, Table 2, p.35.

23. John T Dunlop, "The Limits of Legal Compulsion", *Labor Law Journal*, Vol.27, February 1976, p.67.

24. This is based on Fred K Foulkes, "Learning to Live with OSHA", *Harvard Business Review*, November-December 1973, p.59.

25. John Mendeloff, *Regulating Safety: An Economic and Political Analysis of Occupational Health and Safety Policy*, MIT Press, Cambridge, Mass. 1979, p.154-7.

26. W Kip Viscusi, "The Impact of Occupational Safety and Health Regulation", *The Bell Journal of Economics*, Vol.10, No.1, Spring 1979.

27. Robert S Smith, "The Impact of OSHA Inspections on Manufacturing Injury Rates", *Journal of Human Resources*, 1979.

28. Mendeloff, *op.cit*., p.164.5.

29. Mendeloff, *ibid*., p.165

30. *Challenge*, November-December 1979, p.35-7.

31. Albert L Nichols and Richard Zeckhauser, "Government Comes to the Workplace: An Assessment of OSHA", *Public Interest*, Fall 1977, p.61-2.

32. Nichols and Zeckhauser, *op.cit*., p.62-3.

33. Mendeloff, *ibid*. p.167-70.

34. Lawrence S Bacon, *Bargaining for Job Safety and Health*, MIT Press, Cambridge, Mass. 1980, Ch.4.

35. Nicholas Ashford, *Crisis in the Workplace: Occupational Disease and Injury*, MIT Press, Cambridge, Mass, 1976, p.493.

36. Thomas A Kochan, *Collective Bargaining and Industrial Relations*, Irwin, Illinois, 1980, p.362.

37. Carl Gersung, *Work Hazards and Industrial*

Conflict, University of Rhode Island Press, Providence, Rhode Island, 1981, p.126.

38. Joseph F Follman Jnr., *The Economics of Industrial Health*, AMACOM, New York, 1978, p.207.

39. Wayne Wendling, "Collective Bargaining and Industrial Safety", *Proceedings of the Industrial Relations Research Association*, Winter, 1977.

40. Gersung, *op.cit.*, p.122.

41. Bracow, *op.cit.*, p.74-76.

42. Joseph B Summa, "Criteria for Safety and Health Arbitration", *Labor Law Journal*, Vol.26, No.6, June 1975, p.374; see also Frank D Ferris, "Resolving Safety Disputes: Work or Walk", *Labor Law Journal*, Vol.26, No.11, November 1975.

43. See Roger B Jacobs, "Employee Resistance to OSHA Standards: Towards a More Reasonable Approach", *Labor Law Journal*, Vol.30, No.4, April 1979.

44. Northrup, Rowan and Perry, *ibid.*, p.256-8.

45. Gersung, *ibid.*, p.132-3.

46. Matt Witt, "Dangerous Substances and the US Worker: Current Practice and Viewpoints", *International Labour Review*, Vol.118, No.2, March-April, 1979, p.165.

47. W Kip Viscusi, "Unions, Labor Market Structure, and the Welfare Implications of the Quality of Work", *Journal of Labor Research*, Vol.1, Spring 1980, Table I, p.186, No.1.

48. Witt, *op.cit.*, p.174-5.

49. Kochan, *loc.cit.*

Chapter Eight

SOME ISSUES FOR THE FUTURE

The purpose of this final chapter is not so much to summarise the major findings contained in previous chapters, but rather to redress, at least to a certain extent, three particular emphases embodied in earlier chapters. The first such emphasis is typified by Chapter 5 where we analysed the attitudes, behaviour and potential impact of a sample of safety representatives using three basic sub-vectors of variables that were entirely _in-plant_ in nature. This sort of approach involves the risk of ignoring, or at least seriously under-estimating, the influence of administrative and legal factors _external_ to the individual workplace on the workings of the safety representative and/or safety committee functions. Accordingly, the extent and nature of some of these external sources of influence are considered here. The second emphasis in our work, which followed from the detailed analysis of industrial accidents in Chapter 2, was that we have very largely discussed the safety representative and safety committee functions in terms of their potential for improving industrial safety, rather than occupational health. However, as our review of the United States experience in the previous chapter has shown, _the_ issue for the future is the establishment of joint initiatives and undertakings to come to grips with the much neglected problem area of occupational health and disease. As a consequence we consider here the emerging union attitudes and strategies in relation to the health, rather than safety, side of this subject area. Thirdly, the title of this book indicates that our basic concern has been very much with the extent and nature of _union_ involvement in workplace health and safety problems. This emphasis seems justified in view of the relatively highly organised nature of the workforce in Britain, but some attention should be given to the issue of employee involvement in the workplace health and safety area in non-union plants, particularly as a number of commentators have, as we saw in Chapter 4, expressed considerable concern about the implications of the safety representative and safety committee regulations for these employees. Accordingly, this is the third substantive area of discussion in this chapter.

The final sub-section of this chapter takes up the question of research strategy in this subject area in the future. The consideration of this question should be seen in the light of the basic rationale of this book which has been to analyse certain larger bodies of data with a view to identifying some of the broader environmental and organisational parameters relevant to an understanding of the operation and ultimate effectiveness of the safety representative and safety committee functions. The basic argument presented here has been that such influences should ideally be identified before conducting small scale case studies designed to provide an in-depth analysis of certain individual issues and relationships. If this book has, even at least partially, achieved this aim then there is likely to be considerable value in putting together a systematic, well conceived programme of case studies in this subject area in the future. This is particularly so as there already appear to be a number of situation or sector specific issues and problems arising in relation to the introduction and operation of the safety representative and safety committee functions in certain plants, industries and sectors. The public sector is particularly prominent in this regard so that the final section of this chapter will briefly outline the relevant situation specific issues and problems in this sector with a view to providing an explicit focus for some future case study work.

The Broader Administrative and Legal Set of Influences

The trade unions have for long expressed considerable scepticism about the effectiveness of the relatively small sized Factory Inspectorate in ensuring adequate employer compliance with workplace health and safety regulations; a scepticism perhaps not unfounded in view of some of the findings presented in Chapter 4. The factory inspectorate is still a relatively small sized body, when compared to the number of establishments registered and potentially subject to inspection, although the inspectorate staff has increased from 595 in April 1973 to 685 in April 1976 and to 881 in November 1979.(1) The implications of the relatively small size of the inspectorate can be gauged from the results of a sample survey in June 1979 of establishments on the Factory Inspectorate register; this sample included some, but by no means all, establishments due to be inspected under the terms and coverage of the 1974 Act. The results of this survey indicated that 77 per cent of the establishments had received their last substantial inspection since January 1975, but that only 50 per cent and 32 per cent of them had received it since January 1977 and January 1978 respectively.(2) In addition to the increased size of the inspectorate over the period 1973-79 there have been a number of changes in its operating procedure, based very largely on the recommendations of the Robens Committee Report. (3) The most fundamental procedural change is that (4)

> Since 1976 the (factory) inspectorate has endeavoured to include in its inspection programme the 25 per cent of factories most meriting inspection, which will, of course, include unsatisfactory establishments inspected in the previous year, and some establishments that have not been inspected for a number of years. This may be compared with the 1973 policy of endeavouring, with variable success, to inspect every workplace subject to the Factories Act, however low the risk, every four years.

The Health and Safety Commission is concerned that the value of this change to a non-random inspection programme, which specifically concentrates on 'blackspot' establishments, may be offset by the present Government's emphasis on the need for public expenditure cuts. The Commission estimated, for example, that a 10 per cent expenditure cut would have the effect of increasing the average interval between factory inspections from the present five years to 7.7 years.(5) The likelihood of having to implement a 6 per cent expenditure cut by 1982 would reduce the budget of the Commission by £2.2 million, which is the equivalent of some 260 jobs. These proposed cuts will certainly check the planned increase in the size of the inspectorate, which was due to reach 1144 by April 1983, a result which has, not surprisingly, been viewed with considerable concern by the Commission. The proposed cuts are seen as most likely to adversely affect the day to day inspection programme, although other aspects of the Commission's work programme are also likely to be affected.

At present an inspector appointed under the Health and Safety at Work Act has the power under S.21 of the Act to serve an enforcement notice on any person contravening a statutory requirement or carrying out activities in a dangerous manner. The enforcement notice may be either a prohibition notice, which is reserved for situations where, in the opinion of the inspector, activities are being carried on which involve or will involve a risk of serious personal injury, or an improvement notice, which is issued where there is a breach of a statutory requirement or there has been such a breach in the past which is likely to be repeated. A prohibition notice may come into operation on issue, i.e. an immediate prohibition notice - or it may allow time for the required changes and modifications to be made, i.e. a deferred prohibition notice. The contravention of any requirement or prohibition imposed by an improvement or prohibition notice carries criminal sanctions of up to £400 on summary conviction, or two years imprisonment on indictment before a jury in the crown court, which are much tougher penalties than have previously existed with regard to the enforcement of health and safety legislation. The numbers of enforcement notices issued by the Factory Inspectorate and Local Authorities inspectors for the years 1975-77 are set out in Table 8.1.

The combined figures for the Factory Inspectorate and Local Authorities indicate a steady increase in the number of enforcement notices issued during this period of time; 7,376 in 1975, 8.937 in 1976 and 10,244 in 1977. The vast majority of these notices have been improvement ones, with this category accounting for 67.2 per cent of all notices issued in 1975, 74.1 per cent in 1976 and 76.7 per cent in 1977. However, in cases where prohibition notices are issued it is much more likely to be an immediate, than a deferred, notice.

An employer can appeal to an industrial tribunal against the issuance of an enforcement notice. Such appeals can be made on grounds such as the following:(6) the inspector has misunderstood or abused his powers under the Act; the inspector was

Table 8.1: Number of Enforcement Notices Issued, by Type of Notice, 1975-77

	Improvement			Deferred Prohibition			Immediate Prohibition		
	1975	1967	1977	1975	1976	1977	1975	1976	1977
Factory Inspectorate	4189	4123	4833	670	526	371	1538	1451	1556
Local Authorities	767	2497	3020	132	136	69	80	204	395

Sources: Health and Safety Executive, Health and Safety Statistics 1976, HMSO, London, 1979, Table 2.2 and Health and Safety Executive, Health and Safety, Manufacturing and Service Industries 1977, HMSO, London, 1978, Table 7.

wrong in his interpretation of the law; the breach is admitted but the remedy proposed is neither technically feasible or practicable, or needs more time to be implemented, or is too costly; the breach is admitted but is held to be so trivial that the notice should be cancelled. In practice there have been relatively few employer appeals against enforcement notices, and even fewer successful appeals. In the period to September 1977, for example, more than 20,000 enforcement notices were issued, but little more than 100 appeals were heard by industrial tribunals with the number of successful appeals, resulting in the cancellation of the notice being less than 10.(7) In a number of cases where the notice has been upheld the tribunal has, however, extended the time period for the employer to make the necessary changes or modifications. A fuller indication of the outcome of the issuance of enforcement notices is provided by the figures set out in Table 8.2.

Table 8.2

The Results of the Issuance of Enforcement Notices by the Factory Inspectorate, 1976

Number of firms against whom notices issued	4,497
Improvement notices	3,779
Deferred prohibition notices	514
Immediate prohibition notices	1,236
Full compliance	4,442
Compliance with extension by inspector	772
Compliance with extension by tribunal	7
Compliance through notice being withdrawn by inspector	66
Non-compliance - withdrawn by tribunal	2
Non-compliance - prosecution	95
Non-compliance - no further action	145

Source: Health and Safety Executive, Health and Safety Statistics, 1976, HMSO, London, 1979, Table 3.11.

The contents of the above table indicate that the outcome was full compliance in 80.3 per cent of the total number of 5,529 enforcement notices issued with a further 15.3 per cent resulting in a somewhat lesser degree of compliance. The historical reluctance of the inspectorate to prosecute offenders was a matter that we discussed in Chapter 4.

Although we have no prosecution rate statistics to indicate whether there have been any significant changes in this regard in recent years Table 8.3 below does provide some details on the total number of prosecutions initiated by the Factory Inspectorate for the years 1976-77.

Table 8.3
Total Prosecutions Initiated by the Factory Inspectorate, and Their Outcomes, for the Years 1976-77

	1976	1977
Charges	2,174	2,814
Convictions	2,010	2,546
Withdrawn	97	198
Dismissed	67	70
Average penalty per charge laid	£87	£95

Source: Health and Safety Executive, *Health and Safety, Manufacturing and Service Industries 1976*, HMSO, London, 1978, Table 5; Health and Safety Executive, *Health and Safety, Manufacturing and Service Industries 1977*, HMSO, London, 1978, Table 5.

The largest, single category of offence for which proceedings were initiated was 'matters connected with the fencing and construction of machinery'. The relevant figures under this particular heading of offence were 661 and 959 in 1976 and 1977 respectively. The figures in the above Table indicate a relatively high conviction rate, but considerable criticism has been expressed, by both the Minister of State for Employment and the Chairman of the Health and Safety Commission, about the relatively low level of fines imposed by the courts.(8) The particular concern expressed in this regard was that such relatively small sized fines will not help to ensure that employers take compliance with health and safety regulations seriously enough.

In principle there is considerable personal liability under the Health and Safety at Work Act 1974, with inspectors being able to bring criminal charges against an individual. In practice, however, relatively few such charges have been brought. Some figures provided in early 1978 revealed that individuals prosecuted under the Act only totalled 165 (159 convictions) in 1975, although there was a rise to 365 (326 convictions) in 1976 and to 273 (238 convictions) in the first half of 1977.(9) The vast majority of these prosecutions and convictions were for offences under Sections 2 and 33 of the 1974 Act. At that time the largest fine imposed on an individual had been £250, but with the average fine on individuals and companies being only £72 and £136 respectively in England (£33 and £125 respectively in Scotland). There had been relatively few prosecutions of senior company personnel and officials, although those that had taken place had attracted a good deal of media publicity.

The potentially greater powers of the inspectorate provided by the 1974 Act, together with certain procedural changes in their mode of operation, which we have discussed here, may make them a potentially important influence on the operation, and ultimate success, of the safety representative and safety committee functions. The Kochan, Dyer and Lipsky study of joint safety committees in the United States, which was discussed in Chapter 6, certainly reported that the unions in their sample plants "... have definitely used OSHA as a source of leverage or pressure to be relied upon when faced with a recalcitrant management. The main effect of the presence of the law and the threat of its enforcement has been to increase the commitment of management to deal with safety and health issues and specifically to reduce its resistance to union efforts at improving safety and health conditions."(10) An effect along these lines may well be present, particularly in relation to the operation of the safety representative function, in Britain. Accordingly, future research could usefully explore the extent and nature of the interaction between safety representatives and the inspectorate, together with the environmental and organisational factors that seem to account for any observed variation in the nature of this relationship. The results of such a study would usefully complement the analysis and findings of Chapter 5 which examined, among other things, the nature of the relationship between safety representatives and their workforce constituencies.

In examining the extent and nature of union involvement in workplace health and safety matters via the safety representative and safety committee functions one must be careful not to ignore the potentially important implications of the Health and Safety at Work Act 1974 for the individual contract of employment. It has been suggested that health and safety issues are likely to increasingly figure in unfair dismissal claims.(11) Individuals who consider that they have been unfairly dismissed are entitled to complain to an industrial tribunal, although ACAS conciliation officers have a statutory duty to try to settle the complaint without the need for a tribunal hearing. In 1979 ACAS received 43,406 cases for individual conciliation, and of the total number of cases dealt within that year 63 per cent were cleared without reference to a tribunal. (12) Some 90 per cent of these individual conciliation applications were concerned with unfair dismissal, with some 63.5 per cent of these being resolved without reference to a tribunal. Unfortunately, we have no data or evidence on the extent to which health and safety issues have figured in the unfair dismissal claims successfully conciliated by ACAS.

However, it is clear that health and safety issues are becoming increasingly prominent in unfair dismissal claims that reach the stage of a hearing before an industrial tribunal. It would seem rather premature at this relatively early stage to try and draw any hard and fast 'lessons' or 'rules' from tribunal decisions in these cases. Accordingly in the remainder of this section we simply try to provide the reader with a feel or flavour for some of the issues and decisions in some of the more prominent cases that have been heard.

A number of these cases have concerned the adequacy of physical working conditions, such as the temperature of the workshop or factory. For example in Anslow v Globe Carpet Co. Ltd. (COIT No. 628/204) (13) it was found that an employee who had stopped work because of unsatisfactory heating at his workplace had been unfairly dismissed, on the grounds that the company had put forward no valid reason in law for his dismissal, and that the employee had not contributed in any way to his dismissal. In Mariner and others v Domestic and Industrial Polythene Ltd. (14) the four women applicants returned home from work when they found that the temperature in their workplace was below the statutory minimum laid down by the Factories Act. However, when they returned to work the following morning they were handed letters of dismissal in view of their 'strike action'. The tribunal found these dismissals to be unfair on the grounds that the employer's failure to provide a reasonable working environment was a breach of an implied term of the contract of employment. Their refusal to work in such conditions was a reasonable response to the employer's breach of contract and as such could not amount to a strike or other form of industrial action. This was because there had been no deliberate withdrawal of labour in these circumstances such as to amount to a breach of contract. The issue of employee health and safety complaints was at the centre of an important ruling by the Employment Appeal Tribunal in the case of British Aircraft Corporation Ltd. v Austin.(15) In this case the employee indicated that she wanted safety goggles which incorporated her spectacles prescription and the matter was put in the hands of the safety officer, but the applicant heard no more about it. The applicant decided that she could not carry on working without these requested safety goggles and therefore resigned. The industrial tribunals finding that she had been constructively dismissed was upheld by the Employment Appeal Tribunal who ruled that there is an unwritten contractual obligation on employers to act reasonably in dealing with matters of safety or complaints about lack of safety which are drawn to their attention by employees. In their view so long as the complaint is not obviously frivolous, a failure by an employer to investigate a safety complaint may in itself amount to a breach of contract by the employer entitling the employee to resign and claim constructive dismissal. The important implications of this decision, together with that in Keys v Shoefayre Ltd.(16) are clearly evident in the following commentary:(17)

> In discussing health and safety at work the contract of employment is rarely mentioned. The employers duty to provide for his employees safety is invariably considered in terms of tort.... The contractual aspect of the duty has tended relatively to atrophy only because an action in tort has procedural and financial advantages over one in contract. Nevertheless, breach of this implied contractual obligation has recently acquired increased significance. This is because if such a term is regarded as fundamental, breach thereof can amount to a repudiation by the employer entitling the employee to terminate the contract. This also counts as a dismissal for the purposes of unfair dismissal under the Employment Protection Consolidation Act 1978, S.55(2)(c). An entire new range of possibilities is thus opened up. Furthermore, a claim for unfair dismissal appears to lie in a different, if not wider, set of situations than would sustain an action

for damages in contract or tort. Such actions are dependent upon proof of loss or injury to the employee, whereas to ground an unfair dismissal claim the employee need only establish a simple repudiatory breach, even though he has suffered no damage thereby.

There have been a number of cases concerning protective clothing and devices. In Rushton v Blue Ribbon Equestrian Group Ltd.(18) the applicant was employed on work that was dusty and was accustomed to wearing a mask. The applicant requested new filters for the mask as the existing filter was dirty, but none were available and the applicant said he would not work until a new one was provided. He was instantly dismissed and claimed that his dismissal was unfair, a claim upheld by the tribunal. In some cases an employee's refusal to observe safety rules, such as the wearing of personal protective devices, have been seen by employers as constituting gross misconduct, thus warranting summary dismissal; the employer's position on this matter has typically been upheld by tribunals as, for example, in Mortimer v V.L. Churchill Ltd.(19) and Frizzell v Flanders.(20) However, in the cases of Mayhew v Anderson (Stoke Newington) Ltd. and Henry v Vauxhall Motors Ltd.(21) the tribunals ruled that it is not always fair to dismiss an employee for persistent refusal to wear protective clothing in that there is an onus on management to investigate the reasons for the refusal and to consider alternative types of protective clothing if any are available.

In the highly controversial case of Lindsay v Dunlop Ltd.(22) the applicant, who worked in the tyre-curing section of an old factory, believed that he risked illness and ill-health due to exposure to fumes and dust which were possibly carcinogenic. As a temporary measure until more far reaching improvements could be made, the union and all of the other employees involved agreed to use masks and resume normal working. The applicant, however, believed the masks to be inadequate protection against the fume and dust exposure and refused to resume work. His continued refusal led to his dismissal. The tribunal took the view that the applicant's concern was genuine, but not necessarily realistic. There was no evidence to show that the masks were or were not adequate protection and in the circumstances the tribunal found that the employer was doing all that was reasonable to improve the situation. The dismissal of the applicant was therefore found to be not unfair. This decision was upheld on appeal to the Employment Appeals Tribunal, although in their ruling they stated that where an employee is dismissed for refusing to work in what he regards as unhealthy conditions, it is not the function of an industrial tribunal in an unfair dismissal case to make findings as to whether the employer was in breach of common law or statutory safety duties.(23)

This ruling should be seen in the light of decisions such as that in Howard v Overdale Engineering Co. Ltd.(24) where the applicant claimed that it was unfair to dismiss him for refusing to work in his employer's new factory where a certain amount of dust was generated by engineers drilling cable lines in the floor, thus exposing him to a health risk. The tribunal held the dismissal to be fair as there was no proof of any breach by the employer of a statutory duty to prevent impurities from getting into the air. The necessity to show a breach or likely breach of a statutory safety duty, which was evident in this case, obviously places a considerable burden on employees in trying to establish unfair dismissal in such circumstances. In general, however, the employer should warn employees of the consequences of refusal to work in such circumstances, although in this particular case it was felt by the tribunal that no warning would have had any effect given the individual employee concerned.

(The tribunals appear to have taken a generally 'tough line' with employees who claim that they were unfairly dismissed following incidents involving a failure to adhere to safety procedures. This was certainly the case in Ashworth v John Needham & Sons Ltd.(25) where the applicant was summarily dismissed in accordance with the company's disciplinary procedure which listed 'flagrant disregard of safety duties' as gross misconduct, for failing to replace the proper fencing around a 20-foot deep hole. Similarly in Martin v Yorkshire Imperial Metals Ltd. (26) the applicant was dismissed for making an adjustment to his machine which meant that it could operate without the safety lever being in position. Although some concern was expressed over whether the applicant had been adequately trained in the safety rules and procedures the tribunal held the dismissal to be fair.) A particularly important decision in this area was that in Taylor v Alidair Ltd.(27) where an industrial tribunal held that it was unfair to dismiss the applicant, who was a pilot, on the basis of a single error of judgement. This decision however, was reversed by the Employment Appeal Tribunal who concluded that "In our judgement there are activities in which the degree of professional skill which must be required is so high, and the potential consequences of the smallest departure of that high standard are so serious, that one failure to perform in accordance with those standards is enough to justify dismissal."(28) This finding by the Employment Appeal Tribunal was subsequently upheld by the Court of Appeal. This judgement is likely to be of considerable influence on industrial tribunals assessing whether or not it is fair to dismiss an employee for a serious breach of safety rules in circumstances where the employee has received no previous warnings on this count. The implication of this judgement would seem to be that tribunals should in assessing fairness consider the potential seriousness of the consequences of the breach of safety rules.

One final decision of some interest was that in Chant v Aquaboats Ltd.(29) The applicant had complained to management on behalf of his fellow employees that certain machines were not up to the safety standard required by the Woodworking Machinery Regulations 1974. The machines were in fact below the required standard, and when the applicant received no satisfactory reply from management he suggested to his fellow employees who were not union members that they should join the union. Furthermore, he organised a petition of complaint, which was vetted by his union office before being presented to management. The presentation of this petition led to his dismissal, which he claimed was for reasons of trade union activity. The industrial tribunal held that his actions did not fall within the category of trade union activities, a decision subsequently upheld by the Employment Appeal Tribunal. In the view of the Employment Appeal Tribunal the fact that some of the employees who made the complaint happened to be union members, and that their spokesman happened to be one also, did not make their representations the activities of a trade union. The applicant was not a shop steward and the fact that the union office had vetted the petition did not make it a union communication. The implication of this case appears to be that in the absence of some 'institutional' link between the union and the employer, such safety complaint activities will not be classed as union activities in law.

Union Involvement and Occupational Health

In this section we seek to broaden our perspective on union involvement in the area of workplace health and safety by explicitly considering some dimensions of the occupational health problem, as opposed to our previous concentration on the potential of the safety representative and safety committee functions for improving industrial safety. The first step in considering some of the dimensions of occupational ill-health and disease in Britain involves distinguishing between diseases which should be notified so that the relevant inspectorate can carry out investigations with a view to preventing a recurrence, and diseases which are prescribed for the purpose of compensating individuals who have contracted illness through their employment. There are some 50 prescribed diseases and 16 notifiable diseases, with most, but not all, of the latter appearing on the prescribed list. This 'mismatch' derives from essentially historical reasons, namely the different purposes of the two sets of disease classification.

A list of notifiable diseases was originally laid down in the 1895 Factories Act, but with subsequent regulations extending the number to the present total of 16. In Table 8.4 we set out the relevant figures on the number of notified cases of industrial disease in recent years.

These reported cases of notifiable diseases should be seen in the light of the widely acknowledged deficiencies in the existing procedures for the notification of disease. In this regard it has been claimed that the current procedures involve unnecessary duplication, the provision of relatively little useable information and meagre penalties for non-compliance, with the result that little real attempt has been made to seriously enforce the notification requirements. The Health and Safety Commission has in fact produced a discussion document on the notification of diseases and ill-health, which complements an earlier document on the updating and extension of requirements for the notification of accidents and dangerous occurrences. The discussion document on diseases and ill-health suggests the following categories of notification requirements: (30) (i) communicable diseases; (ii) acute ill-health attributable to incidents identifiable by employers; (iii) occupational ill-health causing more than three days sickness designated by the Dep-

Table 8.4: Number of Notified Cases of Industrial Disease, 1972-76, All Factories Act Premises

Disease	1972	1973	1974	1975	1976
Aniline poisoning	15	12	9	30	35
Anthrax	-	1	2	1	3
Arsenical poisoning	-	-	2	8	-
Beryllium poisoning	1	1	2	1	1(1)
Cadmium poisoning	7	5	2	3	7
Carbon disulphide poisoning	-	-	-	-	-
Chrome ulceration	130	117	71	69	65
Chronic benzene poisoning	-	1(1)	-	-	-
Compressed air	-	2	1	16	34
Epitheliomatous ulceration	36(1)	14(1)	12(2)	11	7
Lead poisoning	82	59	36	27	31
Manganese poisoning	-	-	1(1)	-	1
Mercurial poisoning	-	-	4	-	-
Phosphorous poisoning	3	2	1	-	1
Toxic anaemia	-	-	-	-	-
Toxic jaundice	-	-	-	-	-
All Industrial Diseases	274(1)	214(2)	143(3)	166(-)	185(1)

Source: Health and Safety Executive, Health and Safety Statistics 1976, HMSO, London, 1979, Table 10.1, p.50. (Figures in brackets denote fatalities)

artment of Health and Social Security as a prescribed disease; and (iv) other cases of occupational ill-health.

The concept of diseases prescribed for the purpose of obtaining compensation was introduced by the National Insurance Act of 1948. Under the National Insurance Acts such compensation is an employee right separate from any rights to claim damages under the law of torts. The rate of benefit for the prescribed disease is the same as that for an accident at work, i.e. the industrial injury benefit rate. The current list of prescribed diseases is set out in Table 8.5.

A number of individual unions have long pressed for additions to the list of prescribed diseases. For example, sections of the NUM have pressed for the prescription of emphysema to be added to the already prescribed pneumoconiosis, while the AUEW (foundry section) have been prominent in the recent campaign to have vibration diseases prescribed for the purpose of compensation. The unions concerned were highly criticial of the failure of the Pearson Committee to recommend that these particular diseases, plus bronchitis and tenosynoritis, be added to the list of prescribed diseases.(31) In addition to this long-standing strategy of attempting to add to the list of prescribed diseases for purposes of compensation entitlement there are instances of new union initiatives being taken in relation to problems of occupational ill health and disease. A number of examples may usefully illustrate the sort of actions that are being taken in this regard.(32) The ASTMS has produced a policy statement on visual display unit operations, together with a policy on stress at work and the ways in which unions can negotiate collectively on this issue. The GMWU has run a series of seminars for its safety representatives on identification of substances likely to cause cancer. The GMWU is also planning to run a campaign to secure a lower level of radiation standards than that recently recommended by the International Committee for Radiological Protection. The GMWU is also demanding more stringent controls on the use of man-made mineral fibres, including glass and rock wool, which it considers to be as dangerous as asbestos. To this end data sheets have been produced by the GMWU for safety representatives to send to manufacturers of dangerous substances (especially chemicals) asking for information on their properties.

In the printing industry the unions concerned

Table 8.5

List of Prescribed Diseases

1. Poisoning by lead or a compound of lead.
2. Poisoning by manganese or a compound of manganese.
3. Poisoning by phosphorous or phosphine or poisoning due to the anti-cholinesterase action of organic phosphorous compounds.
4. Poisoning by arsenic or a compound of arsenic.
5. Poisoning by mercury or a compound of mercury.
6. Poisoning by carbon bisulphide.
7. Poisoning by benzene or a homologue.
8. Poisoning by a nitro- or amino- or chloro-derivative of benzene or of a homologue of benzene, or poisoning by nitrochlorbenzine.
9. Poisoning by dinitrophenol or a homologue or by substituted dinitrophenols or by the salts of such substances.
10. Poisoning by tetrachloroethane.
11. Poisoning by tri-cresyl phosphate.
12. Poisoning by tri-phenyl phosphate.
13. Poisoning by diethylene dioxide (dioxan).
14. Poisoning by methyl.
15. Poisoning by chlorinated naphthalene.
16. Poisoning by nickel carbonyl.
17. Poisoning by nitrous fumes.
18. Poisoning by gonioma kamassi (African boxwood).
19. Anthrax
20. Glanders
21. (a) Infection by leptospira icterohaemorrhagiae;
    (b) Infection by leptospira canicola.
22. Ankylostomiasis.
23. (a) Dystrophy of the cornea (including ulceration of the corneal surface) of the eye;
    (b) Localised new growth of the skin, papillomatous of keratotic;
    (c) Squamous-celled carcinoma of the skin. Due in any case to arsenic, tar, pitch, bitumen, mineral oil (including paraffin), soot or any compound product (including quinone or hydroquinone), or residue of any of these substances.
24. (a) Chrome ulceration
    (b) Inflammation or ulceration of the skin or of the mucous membrane of the upper respiratory passages or mouth produced by dust, liquid or vapour (including the condition known as chlor-acne but excluding chrome ulceration).
25. Inflammation, ulceration or malignant disease of the skin or subcutaneous tissues or of the bones, or blood dyscrasia, or cataract, due to electro-

Table 8.5, cont'd

magnetic radiations (other than radiant heat), or to ionising particles.

26. Heat cataract.
27. Decompression sickness.
28. Cramp of the hand or forearm due to repetitive movements.
31. Subcutaneous cellulitis of the hand (beat hand).
32. Bursitis or subcutaneous cellulitis arising at or about the knee due to severe or prolonged external friction or pressure at or about the knee.
33. Bursitis or subcutaneous cellulitis arising at or about the elbow due to severe or prolonged external friction or pressure at or about the elbow.
34. Traumatic inflammation of the tendons of the hand or forearm or of the associated tendon sheaths.
35. Miner's nystagmus.
36. Poisoning by beryllium or a compound of beryllium.
37. (a) Carcinoma of the mucous membrane of the nose or associated air sinuses;
    (b) Primary carcinoma of a bronchus or of a lung.
38. Tuberculosis.
39. Primary neoplasm of the epithelial lining of the urinary bladder (papilloma of the bladder), or of the renal pelvis or of the ureter or of the urethra.
40. Poisoning by cadmium.
41. Inflammation or ulceration of the mucous membrane of the upper respiratory passages or mouth produced by dust, liquid or vapour.
42. Non-infective dermatitis of external origin (including chrome ulceration of the skin but excluding dermatitis due to ionising particles or electro-magnetic radiations other than radiant heat).
43. Pulmonary disease due to the inhalation of the dust of mouldy hay or other mouldy vegetable produce.
44. Primary malignant neoplasm of the mesothelium (diffuse mesothelioma) of the pleura or of the peritoneum.
45. Adeno-carcinoma of the nasal cavity or associated air sinuses.
46. Infection by brucella abortus.
47. Poisoning by acrylamide monomer.
48. Substantial permanent sensorineural hearing loss due to occupational noise amounting to at least

Table 8.5, cont'd

50dB in the better ear.

49. Viral hepatitis.
50. (a) Angiosarcoma of the liver;
    (b) Osteolysis of the terminal phalanges of the fingers.

Source: Industrial Relations Review and Report, No. 175, May 1978.

---

have sought, as a priority issue, information from employers and manufacturers on the properties of chemicals used by print workers. The phasing out of toluene di-isocyanate (TDI) is a priority target as TDI, which is used in making inks and adhesives, can cause asthma and other chest problems. The unions believe alternatives can be found and would like to see TDI phased out completely. Agreements on this have already been reached at Metal Box and Mardon Flexible Packaging. In the printing industry more generally, however, year-long talks with the British Printing Industries Federation and the Flexible Packaging Association concerning the NGA's demands that (i) a target date be set for the phasing out of TDI in printing inks and adhesives, (ii) a code of practice be agreed for the interim use of TDI and (iii) maintenance of earnings for any member sensitised to the chemical, failed to reach agreement. As a consequence the union has instructed its members not to work with printing inks or adhesives which contain TDI.(33) The printing industry unions are also hoping to undertake a study of carcinogenic substances to see if, as has been suggested, there is an especially high incidence of cancer in the industry.

In their desire to address the problems of exposure to carcinogens and other health hazards in the workplace unions should not lose sight of the importance of trying to do something about one long-standing, general 'disease' in industry, namely that of alcoholism. The Scottish Council on Alcoholism has estimated the costs to British industry of alcoholism and alcohol abuse to be in the region of £350-500 million per annum, but relatively few companies have an explicit policy for dealing with this problem, and even fewer have one that involves union co-operation and assistance.(34) Joint union-management initiatives in dealing with problems of alcoholism at the workplace have demonstrated considerable success in the United States, and are very much on the increase there. The Health and Safety Commission has already requested the Medical Advisory Committee to examine the question of alcoholism in British industry and if this external initiative for change is successful one could expect to see an increase in joint problem solving arrangements in relation to alcoholism in Britain in the years to come. The determinants of success of such arrangements are likely to be very similar to those that we have discussed in the case of joint health and safety committees so that the workings of such arrangements should attract the attention of industrial relations researchers, in the future.

At present it would appear that relatively few detailed agreements have been negotiated over health hazards, although there are increasing signs of union activity in relation to health hazards and diseases. This activity is taking a variety of forms such as seeking tighter government standards for exposure to air, dust, toxic chemicals, radiation and other potential carcinogens, promoting the spread of greater information regarding toxic substances found at the workplace as well as the more long-standing approach of seeking additions to the list of prescribed diseases for purposes of compensation. The TUC has already issued guidelines on collective agreements for dealing with health and safety hazards(35) so that one can reasonably expect further union initiatives in relation to ill-health and disease problems, possibly along some of the lines that are being adopted by unions in the United States.

There are a number of indicators of the fact that health and safety matters are becoming of increasing priority and concern to unions in Britain. The more obvious among such indicators would include the following:-

(i) The recent negotiation of a number of detailed safety agreements, such as those in the motor vehicles industry.(36)

(ii) In the mid-1970s the NUM was unique among unions in having a safety department manned by technically qualified staff. Since then, however, the print unions have appointed a safety co-ordinator and ASTMS and the GMWU have now appointed full time safety officers. Other unions have also created posts for safety staff at varying levels and most have substantially expanded their activity on safety issues; for example, almost all the major unions have now produced safety handbooks.(37)

(iii) Following the establishment of the Coventry Health and Safety Movement in March 1976, locally organised union health and safety groups have been set up all over the country; in late 1979 there were at least 22 such groups in existence.(38) These groups have been set up with the following basic aims: to enable union safety representatives to share information and experiences; to provide a library, information and advisory service on health and safety matters; and to provide a focus for local health and safety experts from universities and colleges to contribute lectures on specialised subjects.

Workplace Health and Safety Problems in Non-Union Plants

In looking at such indicators of increased union concern, activity and involvement in this subject area one must not lose sight of the fact that one half of the British workforce is non-unionised, and there still exist substantial inter-industry variations in the extent of union organisation. These facts raise obvious questions about the extent and nature of employee involvement in health and safety matters in non-union establishments. Indeed these questions were given considerable prominence by the numerous adverse comments that followed the repeal, by the Employment Protection Act 1975, of the original provision of the Health and Safety at Work Act 1974 for employee, as opposed to union, appointed safety representatives. Although it was demonstrated in Chapter 1 that union members are disproportionately employed in the high accident rate industries, this finding should not be interpreted to mean that health and safety problems do not exist in non-union establishments. Accordingly, the nature and effectiveness of employee involvement in the area of workplace health and safety in non-union plants is certainly a matter worthy of detailed examination in the future.

The Health and Safety Commission has in fact recommended the establishment of health and safety committees in non-union establishments, although little in the way of practical suggestions were provided for carrying this recommendation into effect. (39) A survey by Industrial Relations Services on employee involvement in health and safety arrangements reported that some 68 (13 per cent) of their returns came from companies with no recognised trade union.(40) In 48 of these 68 companies joint machinery (typically a safety committee) had been established for dealing with health and safety matters. Furthermore, more than half of the non-union respondents had already established arrangements for health and safety inspections, often on a monthly basis. There was also some evidence of health and safety training for employees being provided by a variety of external bodies. The general conclusion of the survey was that the similarities between the composition and status of safety committees in unionised and non-unionised firms were more striking than their differences. This finding is of undoubted interest although it is likely to reflect the fact that the particular non-union firms who subscribe to the Industrial Relations Review and Report are not entirely representative of the population of non-union establishments in Britain. The fact of the matter is that non-union companies are a far from homogeneous group, with studies in the United States(41) suggesting that one can usefully distinguish between the two following groups of non-union firms: (i) the small sized, low standards employer, and (ii) the large sized, better standards employer. The differences between these two types of non-union establishment in matters such as wages, fringe benefits and amenities at the workplace are very considerable. Accordingly, one could well expect significant differences in the extent and nature of workplace health and safety problems in these two, quite separate groups of non-union establishments. And whether these differences in problems are, in turn, reflected in differences in the arrangements for dealing with them is certainly an issue that is worthy of research by industrial relations scholars.

A Final Point on Methodology

The note on which this book ends is one of methodology. The basic concern throughout the book has been to try and identify some of the broader parameters that are likely to be of relevance in studying the operation and ultimate effectiveness of union involvement in the subject area of workplace health and safety. The basic rationale for this approach was the belief that these sorts of factors should ideally be identified before the undertaking of small scale case studies. The strength of the sort of approach adopted here is the representative nature of much of the data analysed, but the major limitation is the lack of detail of much of the data employed. The case study method clearly comes into its own when one reaches the stage of trying to probe individual issues and relationships in an in-depth manner. And this is particularly the case where there is evidence that these individual issues and relationships may have a strong _situation specific_ component to them.

In the subject area of workplace health and safety there appears to be increasing evidence of the emergence of issues and problems that are very much specific to the public sector. This is perhaps not surprising in view of the fact that the Health and Safety at Work Act 1974 extended statutory health and safety coverage to public sector employees for the first time; although even here the enforcement of these regulations faces the much discussed issue of 'crown immunity'.(42) Furthermore, it should be recalled that the original delay in introducing the safety representative and safety committee regulations very largely followed from local authority complaints about the high costs involved. More recently the Association of County Councils asked (unsuccessfully) the Government to relieve it, for the present, of the burden of complying with the requirements of the Health and Safety at Work Act in local authority premises and to permit the appointment of safety representatives and safety committees at the discretion of management.(43) In addition, the regulations providing for the appointment of safety representatives should be seen in the context of an industrial relations sector with relatively little tradition of strong workplace level organisation, as the public sector still remains the bastion of national or industry level collective agreements in Britain.(44) These sorts of issues all seem to suggest that the public sector would be a very useful place for a series of systematically related case studies of union involvement in workplace health and safety matters. These would, of course, only constitute the starting point for a programme of research that should absorb industrial relations scholars in Britain for many years to come.

NOTES

1. _Industrial Relations Review and Report_, No.221, April 1980.
2. _Industrial Relations Review and Report_, No.221, April 1980.
3. _Report of the Committee on Safety and Health at Work, 1970-72 (Robens)_, Cmnd 5034, London 1972,

p.59-71. On the implementation of the recommended changes see *Industrial Relations Review and Report*, No.163, November 1977.

4. *Industrial Relations Review and Report*, No.221, April 1980.

5. *Industrial Relations Review and Report*, No.221, April 1980.

6. This summary is based on the findings reported in the *Industrial Relations Review and Report*, No.169, February 1978.

7. *Industrial Relations Review and Report*, No.169, February 1978.

8. *Industrial Relations Review and Report*, No.135, September 1976.

9. *Industrial Relations Review and Report*, No.171, March 1978.

10. Thomas A Kochan, Lee Dyer and David B Lipsky, *The Effectiveness of Union-Management Safety and Health Committees*, W E Upjohn Institute for Employment Research, Michigan, 1977, p.85.

11. Brenda Barrett, "Occupational Safety and the Contract of Employment", *New Law Journal*, October 13, 1977, p.1011-13.

12. *ACAS Annual Report* 1979, HMSO, London 1980, p.49.

13. *Industrial Relations Review and Report*, No.183, September 1978.

14. *Industrial Relations Review and Report*, No.169, February 1978.

15. *Industrial Relations Review and Report*, No.183, September 1978.

16. Fraser Davidson, "Recent Cases Section: Safety at Work", *Industrial Law Journal*, 1979, p.176-77.

17. (1978) I.R.L.R.476(I.T.).

18. *Industrial Relations Review and Report*, No.201, June 1979.

19. *IDS Brief, No.175*, February 1980, p.10-11.

20. *Industrial Relations Review and Report*, No.203, July 1979.

21. *Industrial Relations Review and Report*, No.167, January 1978.

22. *Industrial Relations Review and Report*, No.209, October 1979.

23. *IDS Brief, No.175*, February 1980, p.11.

24. *Industrial Relations Review and Report*, No.147, March 1977.

25. *Industrial Relations Review and Report*, No.169, February 1978.

26. *Industrial Relations Review and Report*, No.177, June 1978.

27. *Industrial Relations Review and Report*, No.185, October 1978.
28. *Loc.cit.*
29. *Industrial Relations Review and Report*, No.175, May 1978.
30. *Industrial Relations Review and Report*, No.183, September 1978.
31. *Industrial Relations Review and Report*, No.189, December 1978.
32. *Industrial Relations Review and Report*, No.207, September 1979.
33. *Industrial Relations Review and Report*, No.181, August 1978.
34. P B Beaumont, "The Problem of Alcoholism in Industry", *Employee Relations*, 1981.
35. *Industrial Relations Review and Report*, No.153, June 1977.
36. See, for example, *Industrial Relations Review and Report*, No.213, December 1979.
37. *Industrial Relations Review and Report*, No.207, September 1979.
38. *Industrial Relations Review and Report*, No.209, October 1979.
39. *Industrial Relations Review and Report*, No.147, March 1977.
40. *Industrial Relations Review and Report*, No.189, December 1978.
41. See, for example, Daniel Quinn Mills, *Labor-Management Relations*, McGraw-Hill, New York, 1978, Chapter 4.
42. *Industrial Relations Review and Report*, No.161, October 1977; also No.209, October 1979.
43. *Industrial Relations Review and Report*, No.207, September 1979.
44. P B Beaumont, A W J Thomson and M B Gregory, *Bargaining Structures*, Management Decision Monograph, MCB Publications, 1980.

INDEX